THE TRAIL TO KANJIROBA

THE TRAIL
TO KANJIROBA

Rediscovering Earth in an Age of Loss

WILLIAM deBUYS

WITH ILLUSTRATIONS BY
REBECCA GAAL

SEVEN STORIES PRESS
Oakland • New York • London

SEVEN STORIES PRESS
140 Watts Street
New York, NY 10013
www.sevenstories.com

College professors and high school and middle school teachers
may order free examination copies
of Seven Stories Press titles.
To order, visit www.sevenstories.com
or send a fax on school letterhead to (212) 226-1411.

Library of Congress Cataloging-in-Publication Data has been applied for.

Book design by Stewart Cauley and Beth Kessler

Printed in the USA.

9 8 7 6 5 4 3 2 1

Contents

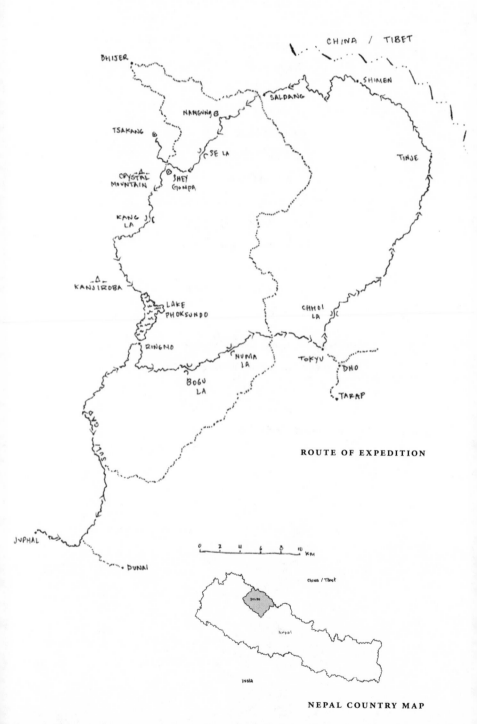

DHIJER

CHINA / TIBET

SHIMEN

SALDANG

NAMGUNG

TSAKANG

SE LA

TINJE

CRYSTAL
MOUNTAIN

SHEY
GOMPA

KANG
LA

KANJIROBA

LAKE
PHOKSUNDO

CHHOI
LA

RINGMO

NUMA
LA

TOKYU

BOGU
LA

DHO

TARAP

ROUTE OF EXPEDITION

SULI

JUPHAL

DUNAI

0 2 4 6 8 10
KM

CHINA / Tibet

DOLPO

Nepal

INDIA

NEPAL COUNTRY MAP

for Nomads past and future

And the fear of you and the dread of you shall be
upon every beast of the earth and upon every fowl of
the air, upon all that moveth upon the earth, and upon
all the fishes of the sea; into your hand they are delivered.
—Genesis 9:2

My genes look out on an ancient landscape . . .
I am a beast born of interaction with environmental
complexity, and to strip me of that complexity is
to render me colorblind, deaf, and tasteless.
——DAN JANZEN

My God is the God of Walkers.
If you walk hard enough, you probably
don't need any other God.
——BRUCE CHATWIN

I am watching ruin, and yet savoring life.
I am complete.
——CHARLES BOWDEN

Introduction

YEARS AGO I wrote a book about climate change, and shortly after that, another book about a beautiful long-horned animal, the saola, one of the rarest large mammals on Earth, which was sliding toward extinction. This book is a continuation of the journey begun by the other two, a further look into dilemmas posed by human transformation of the planet. The subject is fraught with gloomy details, but let me quickly say that those details are not the focus here. This book is about preserving one's sense of joy. It is about finding grace amid the grief.

When the earlier books about climate change and extinction came out, I gave a lot of public presentations. I also talked with fellow writers and researchers. All of us had the too-frequent experience of sensing that our audiences "shut down," that the news from the front lines of environmental change exceeded what people could take in. Of course, the fault may have been our own. Maybe we were not deft enough as storytellers. Even so, the problem of too much bad news remains. As some psychologists put it, people may have "a finite pool of worry," which, once it's filled, has no room for more.

More recently, the world's collective pool of worry has overflowed with urgent concern about the coronavirus pandemic and the economic dislocations it has produced. We have lived amid an onslaught of illness and suffering. None of us can be immune to the temptation to give up or grow numb in the face of nearly continuous bad news. Yet the hard work of Earthcare and social progress must continue. The two are linked and also frequently in conflict. Just treatment of land and wildlife cannot exist without justice for the people who depend on that land. Trouble is, a balance between using the land and protecting it can be exceedingly hard to find. Too often those "balances" result in the continued, albeit slowed, diminishment of wild nature. The hard work of implementing better alternatives and surer strategies never ceases. A further factor has now entered an already complex mix: the world has become better apprised that security from future pandemics cannot be assured without securing stability for the planet's richest reserves of biodiversity. Sixty percent of infectious diseases are zoonotic: they have jumped to humans from other animals. As David Quammen writes, "Ecological disturbance causes diseases to emerge. Shake a tree and things fall out."

At least for me, the situation is paradoxical. My work has taken me to badly disturbed environments where losses are high and prospects for improvement slim. But most of these places have also overflowed with beauty. They have fundamentally changed how I see the world. Studying the climate system has had a similar effect, revealing the complexity of the natural world in new and deeper ways. These experiences have been like arrows from a peculiar kind of Cupid, deepening my awe for the planet.

In the past two centuries our understanding of Earth has approached omniscience. Think of it: we now know how the planet formed, how the continents have moved, how species, including our own, evolved into their present forms. We can finally describe our place in the universe in real—not magical—terms. Certainly there is much that still escapes us, not least in the complexity of ecological systems, and there will always be more to explore, more to learn, such is the inexhaustible variety and intricacy of the world. But the main lineaments of the story that birthed us, the trunk and principal branches of the tree of life to which we belong—these things are known to us with a clarity and depth undreamt by our predecessors. Yet, even as we learned to read the biography of Earth, we have gone far in shredding its most glorious chapter, reducing biodiversity to a degree unknown since the extinction of the dinosaurs. Climate change will accelerate those losses, while producing massive problems of its own. Perhaps the greatest paradox of our time is that, having attained godlike insight, we plow ahead into calamities of our own making.

This book probes that paradox and takes the form of a four-part exploration. One part is physical: in 2016 I joined a five-week, one-hundred-forty-mile medical expedition, in a remote corner of Nepal, hard against the border of Tibet, a land known as Upper Dolpo. Our group, the Nomads Clinic, delivered primary medical care to people who rarely, if ever, see a clinician. I returned to Dolpo with the Nomads Clinic in 2018 for a similar trek in adjacent territory.

A second exploration consists of an intellectual hitchhike with Charles Darwin on his famous voyage of discovery aboard HMS *Beagle*, a journey that inspired him to conceive the theory of natural selection and to launch the greatest revolution in the history of biological thought. Another catching of rides follows an even more meandering route, this one to the theory of plate tectonics, which generated a comparable revolution in the earth sciences. The discoveries that recast the history of Earth in terms of plate tectonics culminated in the 1960s, when I was a boy. Together, plate tectonics and natural selection constitute the discovery of Earth's true past, which must rate among humankind's most extraordinary—and most bittersweet—achievements. It leaves us poised between celebration and sorrow, between success in knowing our place in the miracle of existence and our failure to assent to what we know.

Finally, this book involves a fourth journey, a moral one. We will not "fix" the present wave of extinction or the climate crisis, in the sense of returning to a richer and more stable past. The jig, one might say, is mostly up. So, how to proceed? How to deal with the anger, sorrow, and very real grief that such profound changes produce? I began my journey in the Himalaya wondering if the time had come to apply the ethics of hospice to the care of Earth.

The idea of *hospice for Earth* is easily misunderstood. Of course, Earth is not dying. It has supported life for billions of years, and no matter what we do, our planet will generate life in myriad forms for billions more. But aspects of Earth are passing away. Soon more than an alarmed community of scientists and activists will mourn the lost stability of the Holocene climate, as it becomes clear that a warmer, more energetic, and more turbulent climatic regime has replaced it. Meanwhile, the Sixth Great Extinction, the unmistakable wave of human-caused species loss, which is already underway, will accelerate in proportion to the vehemence of the changes the new climate brings. So, yes, there will be a lot of death, a lot of "patients" breathing their last. And the emotional and spiritual toll on the caregivers who attempt to mitigate those losses and on others who consider themselves family to the ailing world will tap our deepest wells of fortitude.

In Dolpo, the clinics conducted by the Nomads expedition, in villages far from hospitals and laboratories, became theaters in which to observe a particular set of ethics. Most of the time, we could not "fix" our patients—could not put them on long-term drug therapies, let alone intervene with surgeries or other dramatic measures. Instead, the medicine the Nomads clinicians practiced, whether drawing on

Western science or on the traditional healing arts of the Himalaya, was old-fashioned. It emphasized person-to-person contact and relief from immediate suffering. In a phrase, it prioritized *care* over *cure*. Also, it frequently required both patients and practitioners to avoid preoccupation with probabilities over which they had no control—to relinquish their attachment to outcomes—and to focus on the fullness of the present. These ethics, while contrary to the main thrust of Western medicine, are central to hospice and palliative care, and they are far from being defeatist or despairing. As we trekked through the brilliant mountains, the idea of applying such ethics to our troubled planet seemed to me to open possibilities that felt liberating and reenergizing. It offered the prospect of revising one's "terms of engagement" in serving the planet, and this, in turn, raised the possibility of engaging more effectively and with a lighter heart.

First Things

GRAVITY BEING a property of mass, matter attracts matter, and stars and planets form. Within a universe of one hundred billion galaxies, in a galaxy of one hundred billion stars, a nebula takes form, a soup of gas and dust containing the shrapnel of exploded stars. The nebula contracts as matter pulls on matter—mostly single protons, hydrogen atoms. Clots of atoms pull on other clots. A colossus of ever-denser material takes shape. Eventually, the forces at the center of the colossus pack the atoms together under conditions of nearly infinite compression and temperature. Suddenly, a few protons fuse—atoms of hydrogen combining into helium. The reaction releases energy that triggers the fusion of countless protons more. Thus the colossus ignites into a cosmic fountain of heat and light.

In bursting to life, the star—call it the sun—thrusts material outward. Gases travel far. Heavy stuff settles nearer, and the process of gravitational attraction takes hold anew. Bands of small aggregations condense into clusters of larger ones. Orbits cross. Great masses, fiery, molten, and traveling at terrific velocity, crash together, amalgamating into planets. One collision shears off a portion of the sun's third-nearest planet, causing the wreckage to revolve around the body it was split from. It becomes a moon. That collision, as well as others, sets the planet spinning, so that every part of it alternately faces the light of the sun and the darkness of space—day and night. Because the axis of the planet's rotation diverges from the plane of the sun, most of its surface experiences a range of exposures—seasons—as it orbits its parent star.

The universe was already old when these events took place. Time itself was old, having begun when the universe did. Roughly twice as much time passed before the star was born as has passed since then. Although young in relation to most of the universe, the third planet from the sun has already circled its parent more than 4.7 billion times.

It is hard to make sense of a billion of anything. A million is a struggle. A billion, by definition, is exponentially more difficult. A billion minutes add up to 1,903 years, roughly the span of time since the Roman empire reached its height. A billion inches come to 15,783 miles, approximately twice the diameter of Earth. A billion ounces run to 31,250 tons, which is close to the weight of all the buses operated by the Chicago Transit Authority.

Human beings who were anatomically the same as present-day humans came into existence less than three hundred thousand years ago. If the age of the planet were compressed into a day, our species, *Homo sapiens*, would have been around for only the last five and a half seconds. As for the duration of what we call "civilization," you can't push the buttons on a stopwatch fast enough to clock it.

What's astonishing is that we know this. Spinning through space on what Carl Sagan called our "pale blue dot . . . a mote of dust suspended in a sunbeam," we've figured out where we are, what we are, and how we came to be. We are the universe become conscious of itself. It may be that ours is not the only node of self-awareness within the vastness of space—the planets of a hundred billion stars within each of a hundred billion galaxies have the potential to echo, if not replicate, the most particular outcomes—but to say there may be other needles in our cosmic haystack hardly tarnishes the gleam of rareness that is us.

Among our achievements is our success in understanding the history of our planet. We know that in its first billion years, Earth cooled, formed a crust, and then drowned that crust with waters released from magmas rising to the surface. Islands and then continents formed, and their naked geology, slowly eroding, joined with the seas to produce thousands of chemical compounds, eventually including amino acids, sugars, and various fats called lipids. It was still within Earth's first billion years that certain of these compounds, perhaps enriched by residues transported to Earth by meteors, coalesced into combinations that proved capable not just of replicating themselves but of undergoing changes that made their replications progressively more successful. The result, quite literally, was something new under Earth's particular sun, if not under all suns: biological life.

Toward the end of Earth's second billion years, an alga in the planet's caustic seas acquired the trick of photosynthesis, which enabled it to tap the energy of sunlight and metabolize carbon dioxide. As part of the process, the alga emitted a volatile and corrosive waste gas. The alga became so abundant in the seas that, by the early part of the third billion years, this waste product—oxygen—had accumulated to such a degree that it constituted a fair portion of the wreath of gases surrounding the planet. Some of this oxygen compounded into ozone, which had the further effect of shielding the planet from the sun's most intense radiation and allowed the proliferation of still more organisms, all of which, if they had any size at all, were as soft as jelly.

Not until well after Earth's four billionth birthday did hardness in the form of shell, bone, and teeth appear, a development that greatly accelerated the competition among organisms to eat or be eaten. Planlessly and wondrously, ever more complex creatures began to colonize new lands and waters, always in tension with each other and with the resources upon which they depended. Many such creatures flourished for age upon age, but ultimately all of them died away, sometimes in great waves of extinction. The marvel of the planet was that each time the bounty of life contracted, it subsequently rebounded, with new forms arising from the remnants of those that went before.

The most recent flourishing of life on Earth included a proliferation of hairy, warm-blooded animals whose young suckled milk from their mothers—mammals. One of them, a bipedal, large-brained ape, journeyed outward from the savannas where it first appeared and spread across most of the contiguous landmass of the planet. Region by region, the ape's descendants began to differ from one another. Neanderthals occupied the north and west; Denisovans dominated the east; Sapiens abided in a third belt to the south that included their progenitors' original territory. Other, less populous species dwelled elsewhere. The apes embodied a riddle. Possibly, they learned to use an array of tools and, later, developed language because of their large brains, or perhaps the reverse was true—that their proclivity for manipulating things caused their brains to grow big. The cause and the effect may be inseparable. The outcome, however, is known. Tool use and eventually language enabled these apes to occupy their habitats and organize themselves in increasingly complex ways, although different species had different capacities in this regard.

A fundamental feature of the apes' development was that the gift of language intertwined with a broadening consciousness, each teasing onward the other's further growth. Consciousness generated a habit of asking questions about the world. The apes wondered how the world came to be and what forces controlled it. They asked, who made the individual and whence came the family and the clan? Agreement on answers, when agreement was made, helped bind people together, and the apes were indeed people now, for their still-growing consciousness made them different in degree, if not in kind, from the creatures they had been before.

Eventually the Sapiens exhibited an advantage over the Neanderthals, Denisovans, and their other hominid cousins. Perhaps because the complexity of their agreements gave them more cohesion, or because they were smarter, more aggressive, or more adaptable, the Sapiens effected a gradual conquest of the others' territories, killing and breeding their way to dominance until no living trace of their competitors remained except the genes the Sapiens had absorbed in their carnal journey though time.

Probably because of Sapiens' cohesion as hunters, in almost every new territory in which they arrived the largest animals quickly declined. Mastodons and mammoths, long-necked moas and short-faced bears, giant ground sloths and diprotodons, the two-ton wombats of the island continent of Australia: all of them vanished, along with countless others. Sapiens combined intelligence and omnivory to a degree unrivaled by any other predator. They learned to fish, to build boats, to snare birds, to capture creatures of all kinds. Certain animals they tamed and exploited for meat, milk, hide, and fiber, or for tractor power, eventually to pull plows, or for transportation. Other animals, dogs in particular, they accepted as partners and even as members of their households, blurring the question of which species bent the other to its purpose.

Sapiens became expert in the use of plants for both food and medicine. Grasses, especially, opened new possibilities, and when Sapiens learned the trick of intentionally growing such grasses as wheat, barley, maize, and rice—in different places and at different times, but all in the wink of a geological eye—they stumbled into a revolution that fundamentally changed the world. Farming grasses and harvesting their seeds did not make Sapiens stronger or healthier—evidence indicates that it made them less robust—but agriculture enabled more of them to occupy a given area and to feed more children, which meant that ever-larger groups expanded across ever-larger territories, wherever they managed to grow their chosen grasses.

Fed by grains, Sapiens aggregated into villages, which gave rise to towns, which grew to cities, and the unity of each aggregation depended heavily on the persuasiveness of the stories on which the Sapiens agreed.

Typically the stories included accounts of superior beings and invisible forces that shaped the world and alternately favored or hampered the doings of the particular group who believed in them. There is no overstating the importance of the Sapiens' penchant for generating such fictions. By imagining things that had no tangible existence—gods, governments, caste systems, ideas of all kinds, eventually including money—Sapiens developed a degree of cohesion that released immense collective powers. They harnessed rivers to irrigate whole regions; they built immense structures to house their myths; they raised armies and fashioned machines to amplify their strength. And they increased in number, spreading far and wide, always pushing to the margins those of their kind who, having rejected or never known agriculture, persisted in hunting and gathering.

Inevitably, Sapiens began to test their stories against evidence afforded by the actual world, and a new revolution commenced. The habit of evaluating conjectures (hypotheses) against the outcome of actual events (experiments) became a way of knowledge and a culture unto itself. This was science, and it unleashed the explosive growth of both knowledge and capability. In the briefest moment of evolutionary time, Sapiens went from plowing with oxen to walking on the moon, from propitiating demons to mapping the human genome, from believing Earth the center of all things to measuring the breadth of the universe.

In that same moment, the rest of the world changed too. A new great wave of extinction began to sweep through the environments of Earth. Species large and small dwindled to unsustainable numbers and then winked out, felled not just by the prowess of Sapiens as hunters but by changes Sapiens wrought in the land, water, and air. Some of these changes derived from Sapiens' increasingly abundant and complex stream of wastes, which included the gases released by using fire in myriad forms. A few of these gaseous wastes became voluminous enough to alter the atmosphere, much as the oxygen produced by algae had done more than a billion years earlier. This change in the chemistry of the air unleashed chains of cause and effect that Sapiens proved unwilling, if not unable, to halt or control, their genie of invention having long since escaped its bottle.

2

Portrait of the Pilgrim
with Eighty-Eight Toothbrushes

I HAVE a sorrow I cannot shake. It tails me, waits for me around corners, whispers from the shadows. Sometimes it seems to jump into my path. Because of it, I have resolved to go for a very long walk. I am not so naive to think that I can leave this ache of the soul behind, but perhaps by walking I might get ahead of it. Or make it a companion, or at least not an enemy.

And so, September 2016 becomes a month of lists. A list of clothing from hat to boots. A list of gear: collapsible bucket (for washing clothes), pens that write upside down (for making notes while supine in a sleeping bag), headlamp, batteries, repair tape, etc. A list of people with whom I will spend the next six weeks, for I will not be walking alone. Our expedition will include a pulmonologist, a neurologist, a cardiologist, a hospitalist, a hospice internist, a physician's assistant, two ER docs, a naturopath, two Western-trained nurses, a Nepali nursing student, three acupuncturists (German, American, and Nepali), a Western-trained Nepali MD, a traditional Nepali healer or amchi, and a Nepali public health officer. Among the helpers supporting the clinicians will be several Buddhist clerics and three photographers. And me. Our purpose is to deliver health care to the villages of an austere land wedged between the Himalayan crest and the edge of the Tibetan Plateau.

We will be a community in motion, employing scores of mules and horses, numerous guides, camp tenders, and muleteers to transport our gear and ourselves along the precipitous trails of Upper Dolpo, a land identified with Nepal only in modern times and for previous ages part of a shifting assemblage of theocracies and kingdoms. Upper Dolpo is "upper" by reason of both remoteness and altitude, most of it rising twelve thousand feet or more above sea level, with many summits higher than twenty thousand feet and crucial passes, over which we will cross, touching seventeen thousand. A quart of air at such a height contains roughly half the oxygen it has at sea level.

Dolpo lies in northwest Nepal, bordered on the northeast by the Tibetan Autonomous Region, a subdivision of China. (China began its forceful "liberation" of a more truly autonomous Tibet in 1950; the Dalai Lama fled to India in 1959.) The spine of the Himalaya's tallest mountains shields Dolpo from the monsoons of the Indian subcontinent, rendering it a high, cold desert, where success in the brief growing season for barley and buckwheat requires irrigation from mountain streams. Dolpo's communities, which are among the highest-altitude year-round habitations on the planet, are also some of the most delicately poised between sufficiency and want, between satisfied stomachs and empty ones. The Dolpo-pa, its people, are few, widely scattered, and (in Upper Dolpo) ethnically Tibetan. It is easy to think of them as shut off from the rest of humankind and suspended in time, but they too have felt the winds of geopolitical tumult, and they too hear the siren song of the market-driven world that surrounds them like an ocean.

I happen to be the last of the expedition's members to receive an assignment of common goods. In addition to my sleep kit and personal gear, I am charged with delivering to our point of departure a battered duffel containing items that did not fit or were out of sight when other loads were made. In it are a bivouac sack, an air mattress, pills for tapeworm and other parasites; multiple pairs of garishly patterned socks; an assortment of scarves and woolen caps (donated clothing that will be given away); several pounds of dehydrated egg crystals, and similar amounts of dried scallions, freeze-dried berries, pumpkin seeds, and cocoa. The foods are by no means luxuries. We are going to a region that can barely feed itself, let alone a large band of foreigners. The duffel also contains a puzzling bag of several hundred red rubber bracelets bearing in yellow capitals the letters "G.R.A.C.E."; and a supply of eighty-eight toothbrushes, the final complement of fourteen hundred or more stowed in other duffels that are already on their way to Kathmandu. I spent an evening breaking the candy-colored toothbrushes from their plastic packages and grouping them in Ziploc bags. They looked like bundles of party favors. A tall young man of serene disposition delivered these goods to me in the abbot's residence of a Buddhist retreat an hour from where I live. He also entrusted me with thirty crisp hundred-dollar bills, which I carry next to my passport in a pouch under my shirt.

We are a pilgrimage as well as a medical mission, Buddhist in leadership and outline, but including non-Buddhists, like me, and embracing as many purposes as we have pilgrims. My own purpose is to make peace with the sorrow that dogs me. I know it will never leave me alone, for it arises from a circumstance of the world that is as durable as any other. My delight in the beauty of the natural environment must coexist with grief at its destruction. These emotions are like cellmates who cannot get along. They dwell in my head and my heart, and their constant argument creates a moral ache, piercing at times, that frequently sours the taste of life.

The planetary disasters now in constant view—extinction, climate change, and the many ills arising from them—evoke in me a sense of guilt as well as heartache. Much might be said about the failure of my generation (I was born in 1949 and speak from the industrial West) to benefit the world in proportion to the blessings we have received. But viewed from another vantage, the shadow of failure falls on all of us at a species level, if such a thing may be said. Sapiens, on the whole, have proved a flop at tending their nest. We tell ourselves that, for the sake of future generations, we must leave the world at least as good as we found it. And then we don't. Or at least most of us don't. We foul what we have inherited and diminish its beauty. Many of us are hardly sensible of this dereliction, yet I know that the grief I feel is not particular to me. I pray that walking will help.

3

Levels of Amazement

THE FLIGHT attendant, petite and elegant in a tight tunic, passes out cotton balls for us to stuff in our ears against the racket of the turbo-props. We are flying Yeti Airlines in a boxy, workhorse aircraft, bound for the mountain village of Juphal in Dolpo, where a single runway extends the length of a bulldozed ridge. The plane's violent vibrations rattle tray tables and hatch doors, to say nothing of nerves. Southward, out the port windows, cotton-white thunderheads swirl above the overcooked plain of Mother Ganga, the great Ganges River. On the starboard side, the view-warping Plexiglas reveals the Annapurna range, also cloud-wreathed, the snowy mountains erupting from white mists in ominous, pyramidal outlines, like fang rocks on a distant shore.

I am amazed

That this bucket of rivets and sheet metal can stay in the air

That the magnificent Himalaya, Sanskrit's "abode of snow," is
   at hand

That the mountains have arisen from collisions within the
   planet's moving crust

That the uplift has lasted for tens of millions of years, and
   still continues

That we can know this and fly birdlike beside the mountains,
   feeling awe and delight.

Below us lies the zone of intersection between the mountains and the Gangetic plain, which has been the prize of empires since time out of mind. Viewed from eighteen thousand feet, the farmlands and cities bustling with ceaseless activity resemble a gigantic Petri dish, where humans, relentless as bacteria, feed on what they can and spread and multiply. The same might be said of an aerial view of Los Angeles County, or the hinterland of São Paulo or Guangzhou, or countless other regions of the world. Indeed, from far enough away, the entire planet might seem to have become the vessel for a culture—in the laboratory sense—of Sapiens.

The polymath and jack-of-all-sciences Vaclav Smil of the University of Manitoba explains that in 1900, humans and their domestic animals—cattle, sheep, pigs, etc.—represented 83 percent of terrestrial mammalian carbon mass. That last phrase hardly trips off the tongue, but one might consider it to measure the space and energy taken up by the 1.6 billion people then alive. The appropriation of so much animal carbon for humans left only 17 percent for elephants, bears, squirrels, and all other wild mammals.

Smil made a similar calculation for the year 2000, when more than six billion people crowded the planet. He found that humans and their domestic animals had enlarged their share of mammalian mass to 97 percent. Total animal mass—human, domestic, and wild—now exceeded what it had been in 1900 because humans had squeezed more productivity from Earth, but the share belonging to wild things had declined in both absolute and relative terms. Elephants, bears, squirrels, and the rest were down to 3 percent, claiming no more than a sliver in the pie chart of life.

Above the level of bacteria and fungi, Petri Earth is increasingly the domain of a single dominant organism: Sapiens. Human population now exceeds 7.5 billion, with 360,000 more people born every day. By 2050 we will number 9.5 billion, by 2100 perhaps eleven billion. The share possessed by wild things in the bounty of Petri Earth continues to slide toward a flatlined existence only a hiccup north of zero.

There are other ways to depict this trend. The Global Footprint Network calculates that it would take 1.7 Earths to sustain humanity's current hunger for resources and to absorb the resulting wastes. That figure reflects present standards of living. But most people want to improve their lives, and the goal of most economic development is to help them enjoy greater comfort and possess more things. Put simply, large populations from every region of the globe would like to live like Americans or Qataris or the Dutch, which is not good news for the planet. If all Sapiens consumed energy, food, and manufactured goods as voraciously as the average American—and also cast away their wastes, including carbon dioxide, with equal insouciance—the number of Earths required to sustain our planet's present population would rise to nearly five. Embedded within this fact are vast differences in the levels of whim and necessity shaping peoples' consumption. The rich consume more and produce more waste, the poor much less; outside the industrial world, as in Dolpo, the contribution to global pollution of people living at the edge of subsistence is virtually nil. An economic history of the modern age would be required to trace all the causes of our present predicament, but several themes are clear: the so-called developed world, led by the United States, bears preponderant responsibility for the linked calamities of resource depletion and greenhouse-gas pollution. Within most advanced countries, the distribution of the benefits of development has been grotesquely unequal. And many of the governments of advanced nations, especially the United States, have by turns neglected and obstructed efforts to contend with a warming climate, to say nothing of other challenges. Meanwhile, populations burgeon, and people's hunger for food, space, and material goods follows a remorseless mathematics of increase. We are left to realize that not only is Petri Earth crowding out Natural Earth, it is outrunning itself. Its collision with the planet's physical limitations is tectonic in scale, an unstoppable force meeting an immovable barrier. Already, it has pushed up a mountain range of problems, which are rising rapidly to heights that Sapiens have yet to show the resolve or agility to climb.

The plane thrums onward. Making slow progress toward Dolpo, we plunge into bright meadows of cloud tops, past shaggy mountains too steep for a goat. Here and there, a village perches atop a cliff, raw red earth and dark shanties where the forest has been clawed back. Soon a silver river is below us, coursing through a gap in the mountains. We turn northward through the gap and follow the river upstream. As we leave the Gangetic world, the mountains grow taller and more vertical. Freshets spill from the slopes in ribbons of foam. The cliff faces gleam, their rocks hardly weathered. They shed themselves in chutes of boulders and scree, sloughing their geological skin like animals in molt.

4

The Suli Gad

IF THE sky forsook the heavens to incarnate as a river, it would become the Suli Gad, the stream whose headwaters we seek. Its blue waters sing more than roar. Boulders froth the tumbling river into wet clouds, and now, the rain having lifted, scraps of rainbows wink from the haze.

In two days more, having nearly reached the source of the river, we will climb from forests into an almost treeless, lunar world. This will be Upper Dolpo, a land closed to foreigners until the 1980s, when the government of Nepal created Shey-Phoksundo National Park, and opened Dolpo to trekkers, Western pilgrims, do-gooders, and travelers of all kinds.

The declaration of the park drew inspiration from a report filed years earlier by an American naturalist, George Schaller. In 1973 Schaller obtained special permission to enter Dolpo in order to study the bharal, or blue sheep, a high-altitude grazer known only to the Himalaya and the Tibetan Plateau. Schaller wanted to determine where bharal fit in the evolutionary scheme of things—whether they had more affinity with sheep or with goats. To do this he sought to observe a population that humans did not hunt and that therefore behaved entirely according to their natures. He thought he might find such a population in the vicinity of Shey Gompa, an especially sacred place in the inner fastness of Dolpo, where generations of Buddhist lamas had forbidden hunting. Schaller also hoped to glimpse the bharal's fierce predator, the rare and elusive snow leopard. He wrote about his expedition in *Stones of Silence: Journeys in the Himalaya*, which appeared in 1980, but his was not the first or the most famous account of his arduous journey to Shey.

Accompanying Schaller was the writer Peter Matthiessen, whose version of the trip, *The Snow Leopard*, became a nonfiction classic. Most of the book first appeared in two long installments in the *New Yorker*, in March and April 1978. I remember reading the first installment in a state of astonishment, so vividly did Matthiessen evoke the landscapes and villages of the mountains and his own internal struggles. I waited eagerly for the arrival of the second installment, almost praying to the mailbox for its delivery. When it came, I read it without moving from my chair. The story was compelling, yes. But the music of the sentences, the leanness of the writing, and the diamond clarity of the images dazzled me. It was language distilled into a state of being.

Since then, I have reread the book several times, and I have it with me now, as we ascend the same trail (and at the same time of year) that led Matthiessen and Schaller into Upper Dolpo. Or rather, the book has gone ahead of me to tonight's camp, on muleback in my duffel. I relish the passages describing the places we now encounter. Of the Suli Gad and the canyon that enfolds us, Matthiessen wrote, "I wonder if anywhere on earth there is a river more beautiful than the upper Suli Gad in early fall." His musings on vegetation mirror my own. I have noticed wild rose, fringed sage, bracken, and wild parsley, all cousins to species common to the farm where I live in the southern Rocky Mountains. Matthiessen painted the river's canyon as a place of enchantment: "In a copse below Rohagaon, maple, sumac, locust, and wild grape evoke the woods of home, but the trees differ just enough from the familiar ones to make the wood seem dreamlike, a wildwood of children's tales, found again in a soft autumn haze."

5

A Microdot of the Great Rent

OUR TRAIL leads through stands of pine. The trees look much the same as the pines of home. They have the same tall, slender trunks, the same orange-to-black bark, the same three-needle clusters at the tips of their twigs, the same lithe proportions in their limbs. They even appear to have the same problems. Clusters of dead, leafless trunks on the far canyon wall suggest insect outbreaks similar to those brought on by the warming and drying of the environment at home. Not just the human predicament is universal; ecological woes can be too.

The commonest weeds along the trail are also familiar: we boiled nettle for dinner last night, and this afternoon our crew hacked down a field's worth of volunteer marijuana to make room for our tents. Lower down along the trail, thickets of datura, or jimsonweed, choked an empty stone village. It used to be that herders from higher altitudes wintered there when cold weather arrived. No one seems to be sure whether that will happen again this year. Conditions in Dolpo, as everywhere, are changing.

What is sure is that the seeds of trailside weeds hitchhike wherever humans and their animals go. Nettle, marijuana, and datura could have traveled from Asia to North America, or in the reverse direction, by any number of means—their seeds snagged in clothing or animal wool, or accidentally swept into bags of grain. Pines, however, are a different matter. Their distribution around the world and their subdivision into scores of species reaches deep into the planetary past. Two hundred million years ago ancestral pines spread east and west across a supercontinent that geologists have called Laurasia. The subsequent fragmentation of Laurasia into the continents now existing in the northern hemisphere separated the pines of the world into discrete populations. Independently they experienced the many vicissitudes of deep time—isolation, local extirpation, and occasional renewed contact or competition with other populations, to name only a few. Gradually the pines differentiated into distinct species. Two hundred million years is hundreds of times longer than humans have existed—an ocean of time—and yet the pines of the Suli Gad (*Pinus wallichiana*) remain close cousins to the ponderosa pines that stipple the ridges visible from my porch, on the far side of Earth.

Traces of ancient relationships sleep in the land wherever we go. The Himalaya marks part of the ragged edge of Laurasia, where it broke from its parent continent Pangaea. Here beside the Suli Gad, I am standing on a microdot of that great rent.

The southern remnant of Pangaea, called Gondwana, also divided into smaller fragments, each defined by a chunk, or *plate*, of Earth's crust. These became the continents of Africa, Antarctica, Australia, and South America, and the subcontinent of India. Eventually, the Indian plate drifted north and slammed back into a remnant of old Laurasia. The collision, beginning roughly fifty-five million years ago, buckled the edges of both plates, producing the Himalaya. That great welding of continents lies beneath my feet. As the calamitous earthquake that shuddered Nepal in 2015 attests, the welding continues. Geologists estimate that the Himalaya now rises at slightly less than a centimeter per year, roughly equal to the pace at which North America still tears away from Europe along the Mid-Atlantic ridge.

The magmas in Earth's core are forever churning, albeit so slowly as to seem inert in terms of human time. The continents floating above the magmas (because in truth they are lighter) are perpetually in motion. These movements, apart from the sudden shocks of earthquakes, became scientifically detectable only in the last century and a half, and a genuine understanding of them emerged mere decades ago, when those of us now old were children.

6

Circle

UNDER A light drizzle we gather in a corner of the field of hacked-down marijuana where we have camped. Joan Halifax has called us together. She is abbot of the Upaya Zen Center in Santa Fe, New Mexico, and the founder of the Nomads Clinic, which she has led on annual expeditions into remote corners of the Himalaya since 1980. Early Nomads journeys entailed pilgrimages to, or more particularly around, Mount Kailash, a solitary mountain just across Nepal's border with Tibet, where several of Asia's greatest rivers, including the Indus and Brahmaputra, draw their headwaters. The journeys were circum-ambulations of the mountain, a form of pilgrimage known as *kora*. On such trips, the questing Westerners often encountered people in need, and they dispensed aid to the degree they could provide it, more every year. Medical matters have continued to rise in importance to the present time, and our Dolpo itinerary calls for clinics in half a dozen locations. We expect to see hundreds of patients.

But the purpose of pilgrimage remains. Before flying into Dolpo we began our journey with a dawn circumambulation of the Boudha Stupa in Kathmandu, an immense domed temple, which once stood in fields and paddies on the outskirts of the city but has since been swallowed by Kathmandu's sprawl. Many call it the holiest Buddhist shrine in Nepal, and a river of people performing kora flows around it day and night, never ceasing. Our kora consisted of three circuits (everything always in threes) amid the human current, in and out of clouds of sweet incense and the smoke of braziers and butter lamps. We heard the chanting of monks in nearby monasteries, or *gompas*, from which also came the occasional clash of cymbals and the bleat of Tibetan horns. Halifax charged us to use the kora to clarify our intentions for the expedition that lay ahead, but my thoughts were a muddle: did I bring the right gear, have I posed the right questions, will I summon the endurance for five weeks of hard travel? Will I get my over-busy brain to focus on just one thing at a time, or blessedly on nothing at all? I glanced sidelong at my companions. Everyone wore a composed and silent face, but I did not doubt that my anxiety was shared.

Our intended circuit through Dolpo is a kora of sorts, not around a temple or particular sacred mountain, but around a large, wild portion of the district. First we will ascend the Suli Gad to the village of Ringmo, beside the turquoise waters of Lake Phoksundo. Then we bend east through high passes to the populous (for Dolpo) Dho Tarap valley. Over more high country we will trek north to Tinje and Shimen, thence westward to wind-whipped Saldang and beyond to the ancient gompa of Bhijer at Dolpo's northern edge. The names of these places sound to me like words for an anatomy whose structure and function I do not comprehend. My mental picture of them is blank. They are the territory of the unknown.

After Bhijer, we will turn roughly southward to the holy precincts of Shey Gompa, which was likewise the goal of Schaller and Matthiessen in 1973, although they approached from the opposite direction. Leaving Shey we will summit fearsome Kang La, or Snow Pass, which lived up to its name for Schaller and Matthiessen, its deep snowfields almost defeating them. At 17,552 feet Kang La will be the highest and steepest of the passes on our route and our greatest physical trial. Then we will continue south to Ringmo and Phoksundo, completing our circle after more than a month in the arduous beauty of remotest Dolpo.

Halifax drew up the itinerary in consultation with Tenzin Norbu and Prem Dorchi Lama, the operational leaders of our expedition, and also in conversation with the highest lama of the region, Dolpo Tulku Rinpoche, who is a young man still new to the responsibilities of leading his far-flung domain. He will not join us on our travels, but his imprimatur will open doors for us both physical and metaphorical, and he will follow our progress with eager attention.

Everyone addresses Halifax as Roshi, "venerable teacher," a title that recognizes her high standing in the hierarchy of Zen Buddhism. Few women have attained this rank, and she is proud to have broken one of the old guard's more durable glass ceilings. She dresses in black, wears her hair close-cropped, and favors rimless glasses with round, light-sensitive lenses that darken almost to black in the sun. They give her an owlish and vigilant look, as though nothing escapes her attention. At seventy-four, she is the oldest member of the expedition by half a decade. The youngest stands across the circle from her, a wool cap pulled down to his coal-black eyes, earflaps dangling. He is

Vishnu, a lad of no more than thirteen, who watches the world with as feral a look as that of a young Romulus or Remus. Vishnu has come as a late addition among the horsemen who will tend the riding stock. I have counted seventy of us in the circle, Westerners and Nepalis alike. We include horse wranglers, muleteers, cooks, kitchen assistants, Sherpas (the highest-ranking staff, who to varying degrees look after the Westerners),[1] clinicians, and helpers like me. We are a big group, and a lot can happen in a big group. A lot usually does.

We stand hunched against the rain. Sherpas distribute gifts of plastic ponchos and other equipment among the workers, who will keep the expedition fed, sheltered, and moving. Most of the crew hail from Humla, a district northwest of Dolpo, in the very corner of Nepal, close to Kailash. Many are kin in some way to Tenzin or Prem. Tenzin is the proprietor of Sunny Treks, as well as several other businesses, and has outfitted Roshi Halifax's expeditions since 1999. By dint of acumen and enterprise he has made himself one of the wealthiest men in Humla and is surely one of the most capable. Prem, the younger of the two, formerly worked for Tenzin and assisted early Nomads journeys when he was Vishnu's age. He now operates his own trekking company. The relationship between Prem and Tenzin is complicated. They work jointly only on Nomads trips, Halifax insisting that she will not consider an expedition unless both are involved. In this matter, as in most, she gets her way.

The distribution of equipment is consequential. Many of the crew bring little more than the clothes they wear, not for love of traveling light but because they possess little else. The occasional muleteer has been known to go sockless into snow country. Some wear flip-flops on the rockiest trails. By contrast, we Westerners look like fugitives from an outdoor-clothing catalog, clad in shiny Gore-Tex and polypro. In our daypacks we carry cameras, binoculars, and high-tech wind and rain suits, and our duffels, generously limited to twenty kilos (forty-four pounds), bulge with down sleeping bags, inflatable mattresses, and backups for nearly every garment and gadget. We like to imagine we are traveling light, but weights are relative.

After the equipment sharing, we go around the circle repeating our names for all to hear. We also repeat a vow. Most of the names soon

---

1 See A Note on Place-Names and Language, p. 213. The term "Sherpa" is nowadays applied to many guides who are not ethnically Sherpa.

vanish unremembered—there are too many of them, and most of the Nepalis speak shyly, almost inaudibly, maybe because they are unused to speaking in front of strangers, maybe because they don't expect the Westerners to remember their names anyway. The ritual of the vow, however, commands attention. One by one, in Nepali or English, we say, "I vow to take care of myself and to give help to anyone who needs it." Halifax insists on this ceremony. It puts people on notice that the trails ahead can be dangerous and that injury or illness affects the entire community. Several helicopter evacuations have marred previous Nomads treks. But not all emergencies can be defused by an urgent call on a satellite phone and a flight to Kathmandu. One year, falling rocks, loosed by days of rain, knocked a Nomads volunteer, barely conscious, into a fierce, fast river. Her clothes and the current pulled her down. Shedding his pack, a young Sherpa jumped in after her. He reached her quickly and pushed her through the frigid water to the safety of an eddy. She was saved.

He was not. In the act of pushing her forcibly through the wall of the eddy, he necessarily pushed himself away. The current swept him past that point of safety and carried him downstream faster than anyone on shore could run. Witnesses caught sight of his bobbing head three, maybe four times. And then he was gone. They pulled the dazed woman onto the bank. She'd taken a fierce blow to the head, but would recover. Mostly. Her rescuer drowned. His name was Tsering, a name as common in Nepal as John or Mary in America. Many hopeful parents bestow it on their children. It means "long life."

7

There Is No Problem Here

I AM talking with Rigdzin, a man with a full head of bright, almost luminescent, white hair, strong opinions, and shredded, painful fingertips. He is also a man of multiple identities, preferring his Buddhist name to the more conventional appellation on his American passport. Rigdzin is exhausted but does not say so. We have completed our first clinic in the weather-beaten village of Ringmo, twelve thousand feet high, beside the hallucinatory blue waters of Lake Phoksundo. Light rain has resolved to a hanging mist, which is somehow comforting in our state of fatigue. Despite the disorder of a first-time run, we managed to treat ninety-eight patients and sent most of them home feeling better than they had when they arrived. We exhausted multiple boxes of acupuncture needles, dispensed several dozen courses of antibiotics, and doled out toothbrushes, sunglasses, and packets of acetaminophen and ibuprofen by the handful. We also acquired the warm feeling, underlain by unease, that we may actually have done some good.

At this early point in the expedition, everyone projects an earnest and positive attitude. Everyone except Rigdzin. He has more right than most to be cranky. The airline that carried him to Kathmandu lost his duffel, and for a week he has had to make do with only the gear in his daypack. First in Kathmandu and then all the way up the Suli Gad he wore the same soiled T-shirt, which on its back proclaimed in bold block letters,

FEAR
LOVE
SEX
DEATH

The bluntness of the T-shirt matched the bluntness of the man. Rigdzin manages the clinic's pharmacy, and he is mindful of the paradox of our presence here. "Look," he says, "nothing we do here will last. The acupuncture relieves somebody's joint pain for a day, two at most, then it comes back. The antibiotics relieve gastritis for a few weeks or months, a year if you're lucky, then there's reinfection. The old man with congestive heart failure, well, he's going to die. And probably soon."

Rigdzin is repacking the medical duffels for the third or fourth time. Their stiff zippers and the untying and retying of the bags within them have abraded his fingertips, which crack painfully and refuse to heal in the dry, high-altitude air. Rigdzin has been sorting and stowing the medications and medical gear since his arrival in Kathmandu. His approach to his role as master of the pharmacy is obsessive, and it is a role in which obsession can be useful.

In the run-up to our departure, pharmaceutical donations were generous but chaotic. At our first camp, upstream of a hamlet called Rupghat, Rigdzin and several of the clinicians labored over the mass of drugs, wipes, gloves, instruments, and other supplies we had accumulated. They strove to separate the duffels of essential gear from the redundant stores that we did not need. Supplies related to gastric and digestive ailments went into a bag labeled "Gut." "Wounds" got the bandages and sutures. Each of the duffels for "Antibiotics" and "Pain" contained a bushel or two of pills. There were also duffels for "Lung," "Gyn," "Diagnostic," "Acupuncture," and "General," among others. When the sorting was finished, we sent a dozen duffels to the nearby hospital in Dunai, roughly as many as we were keeping. Now Rigdzin fears that amid the confusion, a large stash of antibiotics was mistakenly conveyed to the hospital. We've called by satellite phone to ask that the missing drugs be sent by a direct route to the site of our next clinic, a village distant by several days and two high passes. Unfortunately, the only person at the hospital who can rectify the situation has left for a week-long Hindu festival in the Kathmandu valley, and his return is not expected soon. In the absence of other prospects, Rigdzin persists in searching the duffels that are with us, in case the missing medications have been wrongly stored, not sent away.

He spread the duffels on a tarp near the center of our hillside camp, which lies separate from Ringmo across a sharp defile. Our domed sleeping tents descend the hill like a rash of yellow mushrooms. Below them stand the blue wall tents that are our kitchen and mess hall, behind which spread the startling waters of Lake Phoksundo. The lake strains credulity. You wonder if your eyes tell the truth. The lake, placid beneath towering crags, is improbably cerulean, bluer than the bluest sky and brighter by far than every other color in these sun-dulled heights.

At this altitude, more than two miles above sea level, the unshielded brilliance of the sun mutes the color of things, including the sunbaked faces of the Ringmo villagers, whose clothes in many cases are smudged by the soot of smoky fires. But the lake gleams. If legend told you that a fire had kindled in its depths, you might believe it. Even on a day of mizzling rain, its waters seem incandescent.

Multiple stories explain the origin of the lake. One holds that a demoness, fleeing a powerful adversary, bribed the local villagers with a massive turquoise stone. In return they promised not to tell her pursuer where she had gone. Alas for the demoness—and ultimately the villagers—her enemy was none other than Guru Rinpoche, the spiritual Hercules who banished demons from Tibet and spread the teaching of the Buddha—the dharma—throughout the mountains. Guru Rinpoche saw through the demoness's trick. Aware of the gemstone the villagers were hiding, he turned the turquoise into a turd. Bereft of their reward, they quickly ratted out the demoness, whom Guru Rinpoche thereupon vanquished. Before her demise, however, she avenged herself by drowning the hapless villagers with a lake.

The moral of the story seems to have been lost, perhaps in translation, but the beauty of the lake survives. Also surviving is a Bön monastery on the shore of the lake, which Ringmo has supported for centuries and whose members consider themselves the lake's custodians. No boats ply its waters, and no hunter's gunshots violate its silence. The Bön religion reaches deep into Tibet's demon-riddled past and long preceded Buddhism in these parts. It was in fact the religion that Guru Rinpoche tried to stamp out. It persists here in contradiction to the legend of the lake.

Easier to sort out is Phoksundo's geological history, a hint of which was evident on the long climb up from the Suli Gad. We plodded switchbacks for several hours, up a nearly barren slope, and always the ground underfoot was granular and only randomly rocky. It seemed to possess no structure, no geological identity. We trod on debris, the remains of a pulverized mountainside.

Lake Phoksundo, which gives the appearance of a misplaced fjord, is squeezed between cliffs, filling its mountain chasm to a depth of almost five hundred feet. The natural dam that holds it back came into being when earthquakes loosed at least three immense landslides, causing dozens of cubic kilometers of sandstone and limestone to cascade into Phoksundo gorge. The falling rock disintegrated into near-liquid flows. Much of the detritus converged in the chasm of the Suli Gad, which it filled with rubble. Behind this barrier, a headwaters fork of the Suli Gad, now impounded, formed the lake.

If people lived here at the time of the quakes, the experience would surely have felt demonic. Perhaps the survivors selected the site for their new village of Ringmo because it stands in the open, on a hill of compacted material where nothing more can fall on it. They also did not stint on supplication. Ringmo's squat buildings and low rock walls undergird a forest of prayer flags. The banners fly from poles, parapets, and projections of all kinds. Some of them flutter on freestanding masts tall enough to serve a schooner. The flags are white, the Buddha's color, or they are red, green, blue, yellow, a different message inhering in each color, or they are no color at all, having been bleached by the sun. They cause the dour village, in spite of its earthbound, blocky shapes, to shimmer under the constant wind.

Word of the impending clinic spread with news of our arrival. We were hard to miss, being surely one of the largest groups to have descended upon Ringmo in many years. Rain did not discourage the early arrivals. They queued quietly at the unsheltered triage table, which was set in a gap where the fence of a corral had collapsed. The line of prospective patients formed outside the fence, the clinic inside. Initially the line stretched about fifteen persons deep. Through the day a hundred people came and went, but the line seemed never to grow or shrink. The patients drifted in, waited their turn, and as fast as they were admitted to the corral, others arrived to take their place in line. They explained their ailments at the triage table, where Prem, the chunky, energetic leader of day-to-day operations, entered the patient's name, age, and village in a master logbook. He assigned the patient a number and wrote it with a Sharpie on the patient's hand. He translated the relevant symptoms for Julie, a nurse from San Diego, and Julie checked boxes on a form showing which clinician or succession of clinicians the patient was to see. A runner, a non-medico like me or one of the Sherpas, then escorted the patient to the appropriate location, which might be a blue wall tent, the borrowed stall of a stone barn, or a dim room in the adjacent police quarters. When patients had been delivered to all the clinicians, admissions slowed, and the patients waiting in line collectively sat down on the damp ground. They sat as though filling a long toboggan, each holding the person in front and held by the one behind. They waited without restlessness, in the still, easy way of people used to each other and to a life without furniture, for whom the ground was as good as a sofa.

Most of them were women. We wondered if the men had gone off as laborers to Dubai or Qatar, as many Nepalis have done, but Tenzin told us, no, the women simply work harder than the men, and so they suffer more.

Our patients ranged in age from four to ninety-one. Their faces were round and Mongol, their skin a reddish bronze, their smiles immediate, often gap-toothed. Eddies of laughter moved frequently up and down the line. The women wore nose rings and nose studs, and their ear piercings gaped. Strings of beads hung from their necks, often with a bead of precious coral or turquoise among the wooden ones. Their bodies were rounded and soft, layered in sweaters and thick long skirts, the colors of which approached a brownish monotone. Around their waists they wore aprons of blanket material, which modestly concealed the contour of hip and buttock and completed the uniform of their gender.

Nearly all the women complained of knee, hip, or shoulder pain, ailments honestly earned by toiling up and down the mountain steeps under heavy loads. Vision impairments were also rife, the high-altitude glare of sun and snow being literally blinding, and the barrels of sunglasses and secondhand corrective lenses we brought were soon lightened. Gastric distress, both pain and diarrhea, was also widely shared, and one of our MDs, Geoff Galbraith, using a new, quick diagnostic test, confirmed that the prevailing culprit was a particular bacterium, *H. pylori*, which in the West is frequently associated with stomach ulcers. Other complaints included kidney trouble, respiratory difficulty (too much time in unvented rooms lit by butter lamps and heated by burning yak dung), a leaky heart, and several wounds—in all, a goodly portion of the ills that flesh is heir to. We also offered pregnancy tests and counseling on nutrition and dental care. By the end of the day a feeling of achievement began to nourish our band of clinicians and helpers, myself included.

But Rigdzin, still rummaging in duffels after everyone else left to nap or wash clothes, has more to say: "The situation is absurd. Look around you: our camp equipment is worth more than the annual income of this entire community. We'll go away and pretty much everything here will revert to how it was, including the health of the people we've treated. Follow-up is impossible. It's entropy, imperma-nence, like the Buddha says. Even the mountains and this village will wear away, let alone the little things we try to do."

"So why do them?"

He zips the last duffel and gives a wry smile. "How else are you going to spend your time? Just because in the end it doesn't make any difference—why should that matter? There is no problem here. It's fine. Always will be. We do what we do, and nothing lasts."

8

Dramatis Personae

GEOFF GALBRAITH'S tent is next to mine, and every time I wake, I hear him coughing. It is a racking cough, from deep in the lungs. Geoff is a career hospital administrator lately turned hospice doc. Day by day he looks grayer. He probably picked up a virus on his international flight coming in, and then Kathmandu's intolerable pollution laid him low. He is my age, which puts him among the oldest in our group, and he comes from sea level in Hawaii, which makes the altitude of Dolpo, and perhaps the cold, a serious challenge. He also has scarcely camped before. Every aspect of the expedition is new, from using the latrine to sleeping on the ground, to organizing his gear and dressing for variable weather. Yet he has plunged into a five-week journey from which there is no retreat. He strikes me as the bravest among us. In the morning I ask him if he has started antibiotics. He says he has. His wan smile seems devoid of self-concern. Then he retires to his tent and stays there through breakfast.

Geoff later tells me that he loves medicine but feels something missing, maybe in how it is practiced, maybe in what he brings to it. His unease drew him to hospice work, which led him to take a course from Roshi Halifax called "Being with Dying." The course, together with Roshi's book of the same title, led to his participating in the Nomads Clinic. Other medicos tell similar stories. Wendy Lau, a quiet woman of strong resolve (she is an accomplished kickboxer), is as new to Nomads as Geoff or I. She was born in Hong Kong, attended elite US schools, and fast tracked into the whirl of Silicon Valley. When the dot-com bust in 2000 imposed a pause in her career, she realized she wanted to do more than make gadgets and money. She enrolled in med school. Within a few years, having interned in an ER in New York City, she had seen every kind of human and medical disaster. "So I was fairly burned out even by the time I finished."

The traumas and tragedies weren't the worst of it. She also worked ERs in other high-intensity locations, and in each of them she sensed that the people she tried to serve mistrusted her. They also mistrusted the medicine she provided—they felt "funneled through the system." Under such circumstances she rarely connected at a human level with the people she was treating. "The hard work, the long hours, and the night shifts—all that I can live through." None of it was "as bad as the missing connection." She began searching for an alternative. "I think most physicians when they start med school have an urge to serve and then during the course of learning about medicine and the technicalities and the medical insurance and blah blah blah, it gets shoved away, the service part gets shoved away, and I just really wanted to experience the true service again."

Via the internet, Wendy discovered the Nomads Clinic and the curriculum offered at Upaya. She enrolled in a couple of courses. When she met Halifax, she asked to come to Nepal with Nomads.

"I think we're full," Roshi replied. Then a pause. "But what do you do?"

"I am an emergency medicine physician."

"Oh, well, why don't you go to the next room and fill out the application?"

Half an hour later, Halifax sought her out. "You're in," she said.

Charlie McDonald tells a similar story. His admission to Nomads began in a Bay Area laundromat. He is among the expedition's oldest, like Geoff and me. His irrepressible Will Rogers friendliness and habit of speeding through decisions cause Halifax anxiety, lest he give away everything in the clinic. Years ago, as a long-time pulmonologist in San Francisco, he concluded that he was "doing a lot of service, but not serving." His ten-by-ten examination room and the hospital ICU seemed too small an arena. He took a course on mindfulness because he thought it would help him help his patients, many of whom were contending with terminal illness. He separately started a youth program and built houses and schools in Ethiopia and Haiti. He delved deeper into meditation. Everything helped but nothing felt complete. In the laundromat in Mill Valley, he ran into a Buddhist priest, Arlene. As often happens when listeners truly listen, Charlie spilled his story to her, including his quest for the missing piece that eluded him. Arlene said, "I know what you need to do, and I happen to know there's a space available. And I will call Roshi Halifax today."

That was late 2011 or early 2012. McDonald, a lifelong mountaineer and backpacker, joined the next Nomads expedition. He has returned to Nepal with Nomads every year since. The experience, he says, fundamentally altered his practice of medicine. It changed "how I treat my patients individually, how I meet them, greet them, how I think about them." He now serves as the clinic's medical director, a responsibility he intends soon to share with others, including, he hopes, Nepali members of the team.

Our ranks include Dr. Sonam, who in boyhood worked for Tenzin on early Nomads journeys. He is the first MD to hail from the district of Humla, which is nearly as poor and remote as Dolpo. His degree did not come easily. Not only had he to fight through a bout of tuberculosis in the course of his studies, but the educational obstacles he faced in Nepal sent him far afield to schools in India and Pakistan. He recalls, laughing, that upon his return to Humla after graduation, the welcome he received occasioned so many ceremonial anointments that by the end he literally dripped with yak butter.

The stories of the Nomads clinicians share a theme of restlessness, none more so than that of Wangmo Tsering, a nursing student who hails from the Dho Tarap valley, where we will hold our second clinic. In an airport waiting room at the outset of our trip, I stood next to her and felt an awkward compulsion to make conversation. My questions sounded trite, even to me, but Wangmo put me at ease. Scarcely five feet tall, she seemed uncommonly self-possessed. She said her grandmother raised her, her mother having died of infection following childbirth. The prospects into which she was born were narrow, a future of childbearing, herding, and little else, except that a French alpinist, Marie-Claire Gentric, had come to the valley years earlier. Dolpo enchanted the Frenchwoman, and the extremity of its needs moved her deeply. It happened that she was a woman of means, and she knew how to get more. She organized an NGO, Action Dolpo, and established the Crystal Mountain School, which opened in Tarap in 1994. Wangmo was among its first students and soon learned rudimentary Nepali and English in addition to her native Dopali, a dialect of Tibetan. After seventh grade, the last grade offered at the Dho Tarap campus, the school helped her continue her studies in Kathmandu. And now she is here, amid a cluster of tents beside a surreal lake, conferring with colleagues she never dreamed of having, each with a biography nearly as unusual as her own.

I marvel at the different stories of our clinicians, each similar in its taking of rare opportunities. I also envy them, especially their relationships to their patients. Outside the confessional of a priest, few situations in life permit a more immediate or deeper human connection than the meeting of doctor and patient. When things go right, veils fall away. Trivialities are shelved. Such is the connection Wendy Lau said she sought, a pursuit the others have seconded in their various ways. I think of Anton Chekhov, William Carlos Williams, and Walker Percy, doctors whose closeness to their patients informed their writing. One of our clinicians urged me to read *Cutting for Stone*, a novel by the surgeon Abraham Verghese. The novel hooked me in its first pages, then startled me with an observation that seemed to cast light on the Nomads. Verghese wrote, "Few doctors will admit this, certainly not the young ones, but subconsciously, in entering the profession, we must believe that ministering to others will heal our woundedness. And it can. But it can also deepen the wound."

Verghese's insight does not mean that physicians have more or worse wounds than the rest of us. But they may have a better way of dealing with them, at least most of the time.

9

Uplift

LIGHT RAIN came in the night, bringing snow to the highest elevations. With the morning still gray, we set out for the sky-scraping pass of Bogu La. The map asserts that Bogu La is only a few gasps shy of seventeen thousand feet. Outside a pressurized airplane, many in our group, myself included, have never been so high before. Altitude sickness selects its victims unpredictably, and the pass will be our first real test.

Our trail bends around a shoulder of the mountain, finally narrowing to a footpath chipped from a sheer slope. A thousand feet below, a headwater stream of the Suli Gad boils through a rockbound channel. Our several mule trains have mostly gone ahead, but the last of them, loaded with the remains of camp, catches up to us, and we scramble to perches above the trail to let it go by. The mules bring music to the trail, harnesses jingling, neck bells clanging, each in a different tone. Although ill-fitting saddles gall many a back, the mules are much cherished, and their handlers dress them as richly as their means allow, with ribbons braided into headstalls, and tassels and pom-poms of brightly dyed yak's wool hanging from their necks. Sometimes the lead mule in a gang of five or six sports a little spire of wooly plumes affixed behind the ears, which wobbles and waves as it bobs its head.

We watch the animals dogtrot past, their packs scraping the mountain on one side and jutting over the void on the other. One mule dislodges a fist-size rock from the edge of the trail. The next mule kicks it over the edge. As the seconds pass, I listen hard for the clack of its strike, but no such sound arises from the chasm. There is only the scuffing of hooves, the mule music fading as the train moves on, and the hiss of wind.

High above us, the ridges bear horizontal stripes, alternately light and dark, orderly and even, like the tail of a hawk. I cannot tell if the banding is a feature of the rock or an effect of shadows on stacked ledges. Hours pass, and we climb closer to the realm of the scarps.

Gradually the puzzle resolves. The sun, now high, floods the walls with light, and the banding remains, as obvious as the layers of a cake. White and gray, they run through the height of the cliffs, each band a similar thickness.

The light bands appear to be limestone and the dark ones something else—perhaps dolomite. Such rocks can have originated only in wetlands or saltwater shallows, yet here they pierce the sky eighteen or nineteen thousand feet above their long-vanished maternal ocean, a vertical journey of almost three and a half miles. No less surprising, the bedding of their layers has remained flat. Somehow they accomplished this violent uplift, earthquake by earthquake, without twisting, crumpling, or tilting. Surely the formation of these strata transpired over hundreds of thousands of years, and their upward journey over tens of millions.

It is one thing to entertain big, abstract ideas about the history of Earth but quite another to sense directly the evidence of its biography. I can say to myself, the Himalaya has been sixty million years in the making, but actually seeing former ocean sediments when we have been climbing already for a week—and still to see them high above us—this shakes me. In *Middlemarch*, George Eliot, who was herself an avid amateur geologist, wrote, "There is no general doctrine which is not capable of eating out our morality if unchecked by the deep-seated habit of direct fellow feeling with individual fellow-men." She was commenting on the cruelties wrought by abstract religion, but I like to think that her advice might be applied to Earth. Our ideas about the planet, if we have ideas at all, benefit from lived experience. It is good for our ideas about the world to arise directly from the senses as well as from intellect. It is good for us to receive regular injections of awe.

Charles Darwin set a high example in this regard. In 1831, when he embarked on his nearly five-year voyage aboard the HMS *Beagle*, he knew as much geology as any twenty-two-year-old who had ever lived. The *Beagle* was a small ship with a large crew—seventy-three men crammed into 2,160 square feet of deck space. Darwin had scant room for baggage, but among his few belongings he packed the first volume of Charles Lyell's *Principles of Geology*. It was a gift from his Cambridge mentor John Henslow, who told him to read it but "on no account" to accept its views.

In *Principles*, Lyell argued for "uniformitarianism," the idea that small, incremental changes, like erosion, sedimentation, volcanism, and earthquakes, all of which were observable in the present day, had occurred continuously over immense spans of years and had thus shaped the surface of Earth. The trouble with this notion, which Henslow well appreciated, was that it subverted religion. It suggested that time was longer and deeper than the myths of Genesis and the tedious begats of the Old Testament could account for. To believe in uniformitarianism was to contradict the accepted Christian interpretation of the world. Henslow hoped that no such apostasy would corrupt his student, whom he had favored with an important recommendation. The *Beagle's* mercurial captain, Robert FitzRoy, tended toward depression (and years later would commit suicide). As commanding officer, FitzRoy had to distance himself from the officers below him, yet he also required company and personal interaction to assuage his darker moods. His superior, Captain Francis Beaufort (originator of the Beaufort scale of wind velocities), therefore permitted him to take a companion on the *Beagle's* long journey, someone with whom he might share his table and enjoy intelligent, mentally restorative conversation. It was thought that such a companion might amuse himself with studies of natural history. Ultimately, Henslow was asked to nominate a candidate. He thought young Darwin was just the man.

The second volume of Lyell's *Principles* reached Darwin at Montevideo, Uruguay, in October 1832, nearly eleven months into the voyage. Charles devoured it, as he had the first volume, along with another book Henslow gave him, Alexander von Humboldt's *Personal Narrative of Travels to the Equinoctial Regions of America, During the Years 1799–1804*. At this point in life, had Darwin been pressed to declare what kind of scientist he hoped to be, he might have called himself a geologist.

Young and vigorous, he was fully alive to the world. He absorbed and recorded in his journal impressions of every kind, bad as well as good, which South America supplied in abundance. In Brazil he observed the brutality of Black slavery, a practice his abolitionist upbringing had taught him to despise, and in Argentina he bore witness to campaigns of extermination against the Indians of the pampas. Among the islands at the tip of the continent, he encountered the natives of Tierra del Fuego, whose condition shocked him. To young Darwin their lives seemed unimaginably primitive and debased. These experiences challenged his fundamental beliefs. He began to doubt the solidity of the hypothetical walls of class separating the highest from the lowest human estates. No less importantly, he likewise questioned how wide the gulf might be between humanity and the rest of the animal kingdom. When his shipmates said the people of the archipelago lived like animals, Darwin wondered if there might be a deeper meaning to ponder, beyond the deprecation.

And then, in March 1835, he climbed the Andes.

An earthquake only weeks earlier had awakened Darwin to the power of geologic forces. He was in a forest on the Chilean coast when it hit, and although never in danger, he felt the earth moving under his feet "like a thin crust over a fluid." When he arrived at the port of Concepción, nearer the epicenter, he saw the city's buildings in ruins and learned that many dead had been pulled from the rubble. While exploring the port, he noted beds of shellfish now stranded above the tide line, along with formerly submerged rocks with "marine productions adhering to them . . . cast up high on the beach." He concluded, "There can be no doubt that the land round the bay of Concepción was upraised two or three feet."

Three weeks later, Darwin took temporary leave of the *Beagle* and set out for the high Andes. The winter of the southern hemisphere was approaching, and he took precautions against the danger of being "snowed up" in the heights of Portillo Pass. On the ascent, he noted proofs of "the gradual elevation of the Cordillera," including cobbled terraces in which were embedded the fossils of sea creatures, some of which he collected. The terraces could have formed only underwater. "Daily it is forced home on the mind of the geologist," he concluded, "that nothing, not even the wind that blows, is so unstable as the level of the crust of this earth." A grove of petrified trees particularly amazed him. He encountered the frozen forest on the Argentinian side of the mountains, deducing that the trees had "once waved their branches on the shores of the Atlantic, when that ocean (now driven back 700 miles) came to the foot of the Andes." He saw that the volcanic soil in which the trees had grown had subsequently been overlain by ocean sediments, which meant the fossil trees had first been submerged in the Atlantic and then uplifted to their present position: "I confess I was at first so much astonished, that I could scarcely believe the plainest evidence . . . I now beheld the bed of that ocean, forming a chain of mountains more than seven thousand feet in height."

The Andes produced in Darwin an appreciation of deep geologic time that no mere reading of Lyell could rival. Here were mountains where the evidence of elapsed eons poured in through every sense—in the crunch of graveled seashells underfoot, in the hard stone of petrified trees, and in the taste and smell of the mountain wind. He stood on seafloor far from the sea, amid towering peaks. "Vast, and scarcely comprehensible as such changes must ever appear, yet they have all occurred within a period, recent when compared with the history of the Cordillera; and the Cordillera itself is absolutely modern as compared with many of the fossiliferous strata of Europe and America."

The Andes placed the passage of vast swaths of time before his eyes. They also readied him to encounter the peculiar creatures of the Galapagos Islands, soon to come. The archipelago's biota would puzzle him more than all the mysteries of South America—and would take him longer to decipher. He would unlock their mystery only with an understanding of the phenomenon of time—not ordinary time, but deep time, as ancient as the Andes.

Or the Himalaya. Less than three hundred miles from the pass at Bogu La rises the tallest mountain in the world.[2] British and Indian surveyors calculated its unsurpassed altitude as early as the 1850s. Notwithstanding the perfectly serviceable names bestowed on it by Tibetans and Nepalis (Chomolungma and Sagarmatha, respectively), the British judged the mountain to have "no name intelligible to civilized men" and dubbed it "Everest," after Colonel George Everest, the superintendent of the Great Trigonometrical Survey of India.

Most of Mount Everest's bulk consists of metamorphic rocks, the hardness of which resists erosion and helps account for the mountain's height. Oddly, however, a thin sedimentary formation composes the mountain's utmost summit. This may not seem startling in the present day, now that Earth's restless dynamism is well understood, but in the nineteenth century, such a revelation might have stupefied its hearers. Charles Darwin would have been among the very few humans then alive who would not have been shocked to learn that the tallest mountain on Earth, at 29,031.7 feet, wears a cap of seabottom.

---

2 The meaning of "tallest" depends on the method of measurement. Everest touches the highest altitude above mean sea level. In Ecuador, the summit of Chimborazo, owing to Earth's equatorial bulge, extends a greater distance from the center of the planet. And, measured from seafloor base to cloud-snagging top, Mauna Kea, part of the Big Island of Hawaii, is taller than either of them.

Drift

MANY A ten-year-old, daydreaming in class, has stared at a map of the world and noticed that the eastern bulge of South America, if nudged across the ocean, would fit the curve of Africa's west coast. Other coastlines on either side of the Atlantic promise to snug up nicely too. "I bet this is important," thinks the child. "I hope my teachers will talk about it." I was such a ten-year-old. The year was 1959 or 1960, and in Mr. Magruder's sunlit geography classroom, maps as big as bedsheets hung from the walls. At the time, however, few American geologists, let alone elementary school teachers, had much of an answer for children curious about the fit of the continents.

What Mr. Magruder did not know and could not have told his class of crewcut boys was that a powerful scientific revolution, already underway, would soon reshape human understanding of the history of Earth. It is fair to say that a German geophysicist, Alfred Wegener, launched the revolution with his theory of continental drift, but the advance of Wegener's ideas stalled even before his tragic death in 1930.

Three decades later, the scientists who ultimately confirmed Wegener's insights were not primarily concerned with continental drift. They advocated other novel ideas, but when their theories conformed to Wegener's, they pulled him along in their wake. By the end of the 1960s, Wegener's concept had been reborn as plate tectonics and was soon enshrined as the long-awaited, unifying theory of the earth sciences. The new framework would do more than simply explain the fit of the continents. It would revise, clarify, and reenergize geological thought as powerfully as the theory of natural selection had revolutionized biology a century earlier.

Wegener, born in Berlin in 1880, was an astronomer by education, an Arctic explorer and meteorologist by profession, and a geophysicist by obsession. He was thirty years old, not ten, when a map of the world entranced him with its symmetries. The then prevailing theory of Earth's formation was called "contractionism." It held that, since its formation, Earth had been contracting like a drying apple. As the skin of a shrinking apple becomes folded and uneven, so did the crust of the Earth, creating an irregular topography that had allowed animals and plants to spread via temporary "land bridges" across geographies now separated by oceans. Thus ostriches might be found in Africa, while their obvious relatives, the emus, dwelled in Australia, and still other kin, the rheas, roamed the pampas of Patagonia.

Wegener thought such notions rubbish. He searched for an alternative explanation and examined, not just the mapped coastlines of the continents, but the submerged edges of their continental shelves, which showed an improved fit of one side of the Atlantic to the other—in fact, it was almost perfect. He also probed the evidence of geology and found that from the Cameroons to the Transvaal the rock formations of Africa matched those of Brazil and that the geology of the Scottish Highlands was repeated in the northern Appalachians of America. From paleontology he drew on fossil records to postulate numerous ocean-spanning relationships in distant epochs. He ransacked the evidence of paleoclimatology to discover that the alignment of certain glaciated zones also supported "the fit." And he cited the distribution of living plants and animals, such as ostriches, emus, and rheas, to argue that only the movement of whole continents could account for ancient links between widely separated but related populations.

Wegener first presented his ideas at a geological conference in Germany in 1912, and he continued to develop them even as World War I erupted. Conscripted, Wegener fought on the Western Front and was wounded twice, the second time seriously. During his convalescence he expanded his arguments and produced a book, *The Origin of Continents and Oceans*, which appeared in 1915. Revisions incorporating new data and correcting old errors appeared periodically after the war, including a third edition in 1924 that was translated into English and French, presenting continental drift to an expanded international audience.

The audience, however, was underwhelmed. Wegener faced three massive obstacles, only one of which touched the merits of his case. First, he was not a geologist; his training lay elsewhere, and so the academy shunned him as an amateur. Talented he may have been, but he was an outsider. Second, he was German, and the scientific communities of England and America, which dominated world discourse, were indisposed to heed anyone from a nation that most of the English-speaking world viewed as culpable for the carnage of the Great War.

Some of Wegener's critics used the third obstacle to mask prejudices inherent in the first two, but it was substantial in itself. Wegener argued that the continents had formerly been united in a single great landmass—he coined the name *Pangaea* for his original supercontinent. But his evidence, although voluminous, was circumstantial. He could not convincingly say *how* Pangaea fragmented and the continents moved to their present positions. His theory lacked a mechanism that might account for such dynamism. Some of the ideas he offered, one of which invoked a mysterious force that pushed the continents away from the poles and toward the equator, only opened him to ridicule.

Finally, however, he spied the glimmer of an explanation. In the last edition of his book—in 1929—he cited a hypothesis proffered by Arthur Holmes, a British geologist, which held that magmatic convection deep within the Earth might power movement in the planet's crust. This stunning conceptualization ultimately proved to be correct, but in Wegener's time no empirical measurements to support it could be made. And so the theory of drift languished, neither proven nor disproved, neither accepted nor entirely rejected. Such stasis was also true of drift's rival theories, which included an updated version of contractionism and a third competing view called permanentism, which, as its name implies, argued that the modern continents remained in their original form and position.

Lacking a unifying theory, geology languished as a stepchild of other, "harder" sciences like physics, chemistry, and biology. Most of its practitioners were little bothered by this conceptual void. Economic geologists continued to hunt for minerals and energy reserves, and university professors kept teaching their courses as they had taught them all their careers.

In the course of Wegener's fourth polar expedition in 1930, he and a colleague, Rasmus Villumsen, delivered crucial supplies of food and fuel to a remote camp on the inland icecap of Greenland. Even with the additional stores, the camp lacked resources sufficient for Wegener and Villumsen to stay. They departed via dogsled for another camp, into the teeth of a storm. The deadly weather worsened as temperatures plunged as low as -76°F. The path to safety led through howling wind across miles of broken ice. The journey took Wegener, then fifty years old, to his physical limit, and past it. He died on the trail, perhaps of a heart attack. His partner interred his body in a hasty tomb of snow and ice, which he marked well enough that others later discovered it.

Then Villumsen went on alone, never to be seen again, nor any trace of his body or dog team found.

In the decades after Wegener's death, many researchers added data and new analysis to the case for continental drift. They combed the continents for clues and took the argument as far the evidence of the land permitted. But the continents alone could not tell Earth's story. Crucially, the testimony of the seas, or more particularly, the seafloors, was lacking. It remained for technical advances spurred by World War II and the lavish spending of the Cold War to reveal a universe of facts submerged beneath the oceans. Only then would Wegener be vindicated, and only then might the world begin to know a true biography of Earth.

Bogu La, 16,959 feet

"HEY, DEBUYS! Get out of your tent!" (I think it is Roshi's voice.) "The yaks are coming!"

I crab through the flap, stand, and turn about. In the distance, under a brooding sky, scores of shaggy beasts stream down from the pass. They lurch and glide, their gaits contradictory. They are at once shambling and agile, cumbersome and fluid. Their disordered line flows down the wide, battered trail from Bogu La. The mass of animals must be half a mile long. We hear the whistles and shouts of their herders and the shuffle of hundreds of hooves. The mountain summer has ended, and the herders are taking their yaks to lower pastures.

Most of the Nomads are at the edge of camp, cameras out, eyes rapt on the procession, Roshi among them, a small figure in black with a knitted mulberry hat. We have been friends since back when she was simply Joan Halifax, without the honorific. I cannot remember when we actually met, but we got to know each other through a mutual friend, her last boyfriend as things turned out. He was a hard-drinking poet and former cowboy, who in his youth literally pushed his saddle under the barbed-wire fence at the international border and went off to punch cows in Sonora. Later, thanks to an inheritance, he turned gentleman rancher and conservationist. He was weirdly innocent and profoundly kind, a singular human being who entirely lacked meanness. We commiserated often as he suffered a long decline, and his death brought us still closer together.

Halifax is now halfway through her seventies, and in the world of airplanes, computers, and donors with impressive checkbooks, she moves in rarefied circles. The Dalai Lama and other leading lights are among her friends. She is free to spend her days in the most stimulating and comfortable settings on Earth, but her favorite habitat turns out to be the rockbound trails of the Himalaya and the austere, undefended villages to which they lead.

"Hey, you," she says, still gazing at the flood of yaks bearing down on us, "ever seen anything like that?"

No. Never.

The first yaks veer, almost where we stand, at a great, flat-faced boulder on which long-ago pilgrims etched mantras in intricate script. Glowering, the beasts turn sharply as though in obedience to the carvings, and descend a slope toward the small river that snakes across the floor of the basin. They pass like troops in review, a continuous stream, most of them unburdened, a few bearing the small packs of their herdsmen, their heads low, swaying side to side, the tips of their wide, curved horns scalloping the air, their fiery eyes glaring at us and all the world with suspicion.

Yaks are the great herd animals of the Himalaya. The land formed them as surely as the Great Plains made bison and the Arctic shaped musk oxen. Their hair hangs in wooly skirts, nearly to the ground. Most of the beasts are jet black, a sign of pure pedigree, but some are *dzo*, hybrids of yak and ordinary cattle, sometimes white-faced with woolly coats in multiple shades of brown. The animals grunt as they pass. Taxonomists disagree on the yak's genus, whether to classify them as cattle (*Bos grunniens*) or to separate them into their own class of "grass-eaters" (*Poephagus grunniens*), but they agree that yaks are *grunniens*: they grunt. Because of which, a low drone of bovine complaint forever attends them.

The valley that enfolds us is U shaped, a signature of glaciers past. Waterfalls hang from its cliffs. Tomorrow's trek will take us up the trail the yaks have just descended, bound for a pass my map identifies as Bagala La. The name is redundant, like saying *Rio Grande River*. Hereabouts, "la" means "pass." The mapmakers were evidently not from Dolpo. No one who shapes the information of the larger world is from Dolpo. It turns out that even the root of the printed name is unreliable, for Wangmo has just explained to me that the proper name for the pass is "Bogu," not "Baga."

"Bogu" refers to the spout that carries rainwater past the eaves of a flat roof. The cataracts spilling from the valley's rim resemble the fall of water from such spouts. Hence, Bogu La: *Roof Spout Pass*. The map asserts its highest point to be 5,169 meters above sea level. After Bogu La, we will camp at a trail junction several hundred meters lower and the next day climb to Numa La, which at 5,315 meters, or 17,439 feet, is the highest pass for some distance in any direction. Most of us will walk. Some, including Roshi Joan, will ride horses.

We are close to the roof of the world, and if we are not at the very

top, we are high in the attic. This attic, like most, provides strange
quarters for living. For humans, the key to mastering this environment
was first to master the yak.

During much of the Pleistocene the Tibetan Plateau and the linked
highlands of central Asia lay buried under a blanket of ice. The region
being arid, the Himalayan ice sheet was less thick than that which covered northern North America, but it may nevertheless have averaged a
depth of one hundred fifty meters (five hundred feet). It extended far
down the mountain slopes to modest elevations. When the Pleistocene
epoch ended, two things happened that powerfully shaped the present.
First, the melting of the ice sheet relieved a weight upon the land, providing new impetus to the uplift of the mountains. This unburdening
augmented a period of already furious tectonic activity, so that the
Himalaya rose faster than ever before, gaining thousands of feet in a
few thousand years. By the end of the Pleistocene the Himalaya had
nearly attained the heights that characterize it today.

Second, as the ice retreated, wild yaks followed. They grazed where
grasses and other plants colonized the uncovered land, and the cold of
the snow country suited them well. Their pelage insulated them like
an Arctic goose: a first layer of downy wool cocooned their bodies and
graded out to thick, coarse fibers that shed the rain and snow. Their
stocky build minimized surface area, conserving heat. Their barrel
chests, elongated with one rib more than cattle have, expanded their
lung capacity for the thin air of high altitude. Although yaks prefer
temperatures within a few degrees of freezing, they tolerate conditions
down to forty degrees below zero (which happens to be the same in both
Celsius and Fahrenheit scales). Even today, no one knows the limits of
cold that a yak can endure.

The hunting people who followed wild yak into the mountains may
have brought with them herds of tame yak, domesticated elsewhere,
as they also had sheep and goats, which had been domesticated still
earlier. The nomads herded, hunted, and gathered. The growing of
barley, which came later, spurred the establishment of villages. The
potato, which arrived in the modern era after the discovery of the New
World, nudged the mountain population yet higher.

The early nomads of the mountains used the yak more thoroughly
than even the Plains Indians of North America used bison. They ate its
meat, marrow, and organs. They made tools from its bones, and their
rope, clothing, and shelter from its hair and hide. They also milked their

tame yaks (or, more accurately, their *ghi*, as the female of the species is called), even preserving the milk as a hard, slow-to-dissolve cheese. ("Yak cheese: put in mouth," Tenzin tells me. "After a while, tasty.") Yak butter lamps pushed back the gloom of night. Yak dung, which releases twice the energy of other manures, fueled cooking and heating in the world above timberline.

The heritage of the yak people was Mongol-Tibetan. Unlike the Sioux and Cheyenne, they also used their talismanic animal for transportation and tractor power, burdening their yaks with their belongings as they moved from place to place and ultimately yoking them to plows to till their fields. Without the yak, they could not have sustained themselves in so austere an environment. In the annals of adaptation, no partnership has benefitted humans more than the domestication of the yak.

The yak world and the bison world have many affinities. Even now, in the shadows of Bogu La, the two hundred yaks disperse, not like cattle in bunches clotting the river bottom, but in long straggling lines like bison, drifting along the stony slopes, their bodies dark and bulky. In the evening half-light, as darkness pools beside the river, the silhouettes of the ranging animals recall those of Yellowstone bison in a frontier painting.

We have traded epochs here. By ascending the mountains we have reentered the Pleistocene. Our climate is now the climate of a time when wild yak roamed broadly across Asia. Humans evolved late in that period. They discovered agriculture and built towns and cities still later, after the Pleistocene had passed. Virtually everything that we count as "civilization" is a product of the warmer climate of the last ten or twelve thousand years, a period we call the Holocene. The difference in mean temperature between the Pleistocene and the Holocene is about 4°C (7.2°F). The transition from one to the other was effected over the course of several thousand years. Four degrees of change were enough to transform the planet. Now with human-forced global warming, a new transition of equal magnitude is underway, and it is proceeding, in comparative terms, at lightning speed. If the present rate of warming continues, the difference in mean annual temperature between the middle of the twentieth century and the end of the twenty-first will also be at least 4°C, but it will be accomplished in only a century and a half, a mere heartbeat of geologic time. Absent action more decisive than anything now underway, the planet is on course to depart the Holocene and bid goodbye to a climatic regime whose stability undergirded every major human accomplishment. The new age we seem bound to enter has been dubbed the Anthropocene, the human-shaped epoch, and scholars will long argue whether it properly began with James Watt's steam engine or with the crossing of some other threshold that left its mark in the stratigraphy of rocks and soil. In either case, the course ahead is entirely unmapped and promises to be more turbulent than anything Sapiens have previously experienced.

Powerful heat waves, droughts hotter and more severe than those of the past, and rising seas have already announced themselves. Not to mention supercharged typhoons, hurricanes, floods, blizzards, and other extreme events. Thus it begins.

For reasons incompletely understood, climate-warming proceeds faster at high altitudes than it does lower down. The causes probably include the land's changing albedo, or reflectance, as snowpacks melt and newly bared ground absorbs more heat from the sun. Dust at high altitudes, which darkens the snowpack, and haze below, which shields lowlands from some of the sun's intensity, probably also play a role. It appears that in recent decades lands above four thousand meters (13,123 feet) have warmed 75 percent faster than lands below two thousand meters. Warming is troublesome everywhere, but it is especially problematic where it causes glaciers to recede and alters local hydrology, as the barley farmers of Dolpo are unhappily learning.

The Himalayan past belonged to the yak and its people, but a warming future will not favor the stolid bovine of the Ice Ages. The conditions of high-altitude lands are changing too fast. Yaks guided humans into the attic of the world, but they cannot go through the roof. The passing of the Holocene will leave them homeless. For our band of travelers to have glimpsed them roaming the basin at the foot of Bogu La was a rare and extravagant gift. We watched them as the day dimmed. Their hulking shapes stippled the moraines like chunks of black night fallen early from the sky.

12

On the Trail

IN THIS country, the pass is important, the peak not so much. Summits abound, more than you can count. But useful passes are few. You come to know each of them as a particular place. A pass first opposes you, and then, the crest won, it liberates. The long-time people of the Himalaya revere the passes, tell stories about them, and endow them with powers. They make them places of prayer and thanksgiving, even of rebirth. The general wisdom asserts that every time you cross a pass you start over, you begin life anew. Because the labor of ascent is purifying, the pass gives you a chance to leave the useless parts of yourself behind. And then you cross over into a new world, cleansed and fresh.

It is also said that you should cross a pass early in the afternoon. Demons are everywhere, ready to beset you, but they tend to be lazy. They sleep late and linger over morning tea. Because they are slow to get moving, they will not reach the top of the pass until late in the day. You want to get there ahead of them.

Above all, when you crest the pass you will want to show your thanks that the strain of the climb is over, that the weather has been kind, or if it has not, that it has not been cruel enough to defeat you. You rejoice that you are hale and can stand on an exhilarating knife-edge between two worlds, between the past from which you have climbed and the future into which you will plunge.

As you near the summit of the pass, you hear the cheering of those ahead of you. "Kiki soso lha gha lo," they yell as they reach the *thobo*, the sometimes-enormous pile of stones that marks the top. When you reach the *thobo*, you shout the phrase too, and so do the companions behind you as they straggle in: "Kiki soso lha gha lo!"

There is no exact translation. The phrase has elastic meaning. It means the feeling in your throat when you shout it. It means *JOY!* Or it means everything to do with a pass, from outwitting demons to starting a new life. Everyone grins. There are hugs, handshakes, and claps on the back all around. You dig out your phone or camera and take celebratory photos. A chilling wind keeps your jacket sleeves and the legs of your pants flapping madly; in fact, the flapping of clothes and the roar of wind past the hood of your parka are so loud you must shout to be heard, but no one minds. The hard work of climbing is over. It is time to rejoice. And in a gesture of thanks, you carry a stone to the thobo, which like so many things in this sacramental landscape is said to symbolize the mind of the Buddha. Innumerable travelers before you have built the thobo into a little hill. They have also tied a spider's web of prayer flags to the rocks. The flags snap and fray on wind-taut lines, so that the cairn looks like a sailing ship with its rigging fatally tattered, wrecked on a mountain reef.

At the top of Bogu La, we make our small offerings, and we do the same the next day on Numa La. Some of us tie strings of prayer flags to the existing tangle; some add *katas*, or prayer scarves, bought cheaply by the dozen in Kathmandu. In my tent, at the camp where we saw the yaks, I had written the names of family and close friends on katas, which I leave fluttering on the thobo of Bogu La, soon to unravel in the roaring gusts. The names and the intentions behind them are forms of prayer, which the wind carries off to their destinations. Or to nowhere. I wish I had possessed the foresight to bring a little flag of Earth—bearing the famous Apollo 17 "Blue Marble" shot or something like it—to leave on the thobo amid the fluttering katas. Another prayer for another deserving entity, but I did not think that far ahead.

It seems everyone has brought an errand to Bogu La. I watch as Michael Lobatz, a neurologist from San Diego, solemnly ties a long string of prayer flags into the tangle of offerings. He does it in remembrance of his best friend and business partner, who died of cancer the previous year. The last knots are difficult as his fingers grow numb in the cold, but he persists and slowly, patiently finishes. Then he walks off a distance from the rest of the group and sits on a patch of gravel where the wind has blasted the snow away. I look around for tall, statuesque Julie, the nurse who managed the triage table with Prem at the Ringmo clinic. It was her husband who had died. I don't know how I missed her, but by the time I look back at Michael, Julie is sitting silently beside him.

Then a melody of mule bells breaks through the wind roar, as one of our pack trains arrives. "Kiki soso lha gha lo!" the muleteers shout. Smiles. Pictures. More high fives and slaps on the back.

But soon the cold begins to pierce my inmost layer of clothing. I cast a final glance at the bestrewn thobo and take to the trail, following the jingle of the mules down a long, barren slope. There are miles yet to go, but the trail pitches easily downward, through remnant snow and across patches of tundra and black gravel slides. It crosses gullies scoured by snowmelt torrents and, under a threatening charcoal sky, leads toward a limitless horizon of dun ridges and mazy gray peaks.

Ahead and behind, people walk in small groups, which gradually string out, breaking into threes, twos, and ones. I usually prefer the ones. It is good to walk alone, to match freedom of movement with freedom of thought, for the presence of another person can be constraining, even in silence. One wonders how that person is feeling. Whether he or she is tired, hungry, or wants to rest or talk but doesn't say so lest you do not. Or perhaps you want to pause and just look at a far horizon, or wander into a ravine and take a leak, or walk especially slowly, or rather fast. And what would the other person think about that?

Such thoughts are noise. Solitude is less noisy. There will be plenty of time in camp for conversation, plenty of time to hear everyone's news and enjoy the pleasure of interaction. But in the meantime, the trail is a place to strip down and shed the unnecessary.

At home, it is so often the reverse. Clutter abounds. So much of our lives centers on the things we own and consume. At a societal level, we even tell ourselves to measure well-being (through GDP and similar indices) by the volume of the things we purchase and the rate at which we do so. More is better. Faster is better too. To say otherwise in the political sphere is apostasy. In order to maximize the buzzing of our economic hive, we submit ourselves to a bath of urgings. Signs, screens, and the amplified voices of strangers exhort us to buy food and drink, clothes and contraptions, and to ally with one company to fatten our wallets and another to slim down our waists.

On the trail we towel off from the bath of urgings. The din goes silent, and sounds of place arise. We take mental as well as physical steps. Toward what? The options are profuse. We construct our thoughts from what is before us, and steer our course according to weather, terrain, and other stern conditions. Within those limits, paradoxically, we enjoy great freedom. Every day on the trail becomes a kind of pilgrimage, a *yatra* in the language of the Himalaya. Pilgrims are people who journey toward sacred goals, and what lies nearer the sacred than an understanding of how to live and for what?

Such freedom includes responsibility. On the trail we render assistance to our fellows, should they need it, and strive to stay in good form. We must avoid getting hurt, which, besides being unpleasant, imposes burdens on others. And so we walk alertly, mindful of our step. We find the pace that is right, and keep to it. I will confess to walking like an old man, slower than most, but steady. This is also how a Sherpa walks, although slow for a Sherpa looks fast to me. One keeps moving, which is not a bad metaphor for most situations. I think about this as I age: that the only way to keep moving is to keep moving—a scrap of reasoning no less true for being circular. And so we move down the trail, you and I, with the goal of moving mindfully, aware of our step and aware, too, of our companions ahead and our companions behind and of others whom we may meet along the way, any of whom may need our help, as we may need theirs. Which amounts to a pretty fair metaphor for other, larger journeys.

13

Tokyu Clinic

DAYS AGO, Roshi told me, "My secret plan is to help these Western doctors change their focus from cure to care." The plan may be working.

Michael Lobatz, who recently stepped down from heading the prosperous clinic he founded decades ago in San Diego, has a practice many other doctors would envy. He does things his way, with great success, seeing only the patients whose cases interest him. Not least, he charges plenty, and "they pay in cash." But the clinic we held in Ringmo rocked him: "The poverty hit me like a fist. I've never seen anything like it." That kind of jolt will soften a person, and I sense it is having an effect on Michael, as well as others, including me.

Now he and the rest of us are preparing to meet another surge of ailing and aching people. The prospective patients of our second clinic began to gather in front of the gompa shortly after sunrise.

We are in Tokyu, where two noisy creeks converge to water the long Dho Tarap valley. An app on my on my phone gives our altitude as 13,829 feet. Tokyu is the highest village in Dolpo and one of the highest permanent human settlements in the world. Unfortunately for village farmers, we have come at an inconvenient time. The barley harvest has reached its culmination. Two weeks ago, maybe more, every available man, woman, and child strode into Tokyu's waves of grain armed with sickles to cut the stalks. Stooped over, they crept through the fields hacking the stems a fistful at a time. This was backbreaking labor, as is the local digging of potatoes, which is accomplished with a short-handled hoe. The scythe and the long-handled hoe, if they are known, must be unavailable or unsatisfactory.

After cutting, the harvesters gathered the barley into sheaves, binding each sheaf with a long barley stem tied in a soft knot. They bunched and stacked the sheaves a dozen at a time and set them on end, grain downward, leaning against each other to dry. The brushy pyramids now fill the fields, a yard or two apart, hundreds of them, no, thousands, repeated in every rock-walled plot on both sides of the river. From a distance, they appear to enlace the valley in a web of dotted lines.

At first light the rock walls and the stacks of barley sheaves cast long shadows, the barley tan and the rock walls rust-colored where the early light slants into them. Narrow lanes follow the roll of the land, snaking between the walls. At the end of each lane squats a low dark house, with a curl of smoke rising from its roof.

Suddenly, I see mounds of barley moving. Two great bundles, nearly as big as hayricks, inch up a lane by the river. They seem self-propelled until, with binoculars, I make out a pair of legs at the bottom of each. The bodies belonging to the legs disappear within the bushy loads. Now I see that a man is in front, hands at his shoulders gripping a strap that holds a sagging mass of barley sheaves, and behind him comes a woman equally burdened. I follow their progress to a large walled pen behind a mud-plastered house. "Tashi delek," I say when I reach the wall of the pen. They have deposited their loads and released the carry straps so that the sheaves spill down. "Tashi delek," they answer, a little out of breath.

"Tashi delek" is as much Tibetan as I can say—a common greeting of blessing and good wishes, after which, with no other words in common, we exchange slightly embarrassed smiles. They rest no more than a minute, then draw a blue tarp over the mound of sheaves, coil their straps, and set out again for the fields.

Our clinic is inconvenient because we have appeared at a time when several hours taken from the field to see a doctor, let alone the better part of a day if one travels from a distance, entails risk as well as lost time. The anxious weeks of drying have gone well, but now clouds have begun to gather and it is imperative to get the barley out of the fields and under cover. Not all the farmers are so poor that they haul their harvest on their backs. Some load the grain on horses, some on yaks, and as the sun continues its climb, the fields and lanes quicken with the movement of people and animals.

Soon all of Tokyu is aquiver. Cumulative tons of grain trundle down the lanes, drawn by people and beasts, and no growl of engines fractures the day. Barley is the staff of high-altitude life. Its roasted flour, called *tsampa*, constitutes the entirety of many, if not most, Dolpo meals, prepared as a gruel and flavored with a pinch of salt and a dollop of yak butter or mustard oil. The villagers downstream in Dho, in the neighboring village of Tarap, and in all the hamlets of the valley labor with the same urgency as those in Tokyu to collect their barley and reap the final benefit of the farming year. Even so, more than two hundred of their neighbors have made their way to see the medicos at the clinic of the "round-eyes," as Roshi jokingly calls us.

As in Ringmo, they queue at the triage table where Prem, aided by Julie and later by other clinicians, writes a number on the patient's hand with black marker and assigns him or (more often) her to an examination tent—women's health and pediatrics, general medicine, acupuncture and bodywork—or to Amchi Lhundup or Dr. Sonam, who work at small tables in a recess of the dim gompa, or to the cardiology and neurology specialists, Ranjit and Michael, who started in the gompa but moved outside in search of warmth and light.

Acupuncture patients, scores of them, go first to an unusual feature of the clinic, the foot-washing station. The commonest complaint of the patients is pain in the hip or knee. For treatment, shoes and socks must come off, and on warm days in previous clinics, the smell of unwashed feet corrupted the air within the treatment tent and made working there unbearable. On cold days, with the tent flaps closed, it was worse. Moreover, the sites of needle insertion needed to be clean. These factors mandated the preparatory washing of feet and produced a tableau especially pleasing to Roshi, with its biblical overtones and cultural comedy. Men from our group perform the service, and the patients of the clinic, mostly women, receive it. The sight is arresting: rich Western men kneeling at the feet of Tibetan peasant women. The resulting surprise, embarrassment, and incidental tickling make the foot-washing station a node of laughter and hilarity within the diligent urgency of the clinic.

Runners shuttle patients to and from their proper destinations. They hurry to the gompa to fetch prescriptions from Rigdzin, whose pharmacy occupies a particularly frigid alcove just inside the entry. In the first hours of the clinic I was such a runner, but as the day wore on, Jigme, Pau, and other senior guides joined the ranks of the runners and left me free to enjoy the spectacle.

A hundred people huddle by the triage table before the gompa doors, the line spilling to a terrace below. Old women squat in groups of four and five, no space between them. Compared to Ringmo, the faces of the crowd show less of the Aryan heritage of lowland Nepal and more of the features of Tibet and its Mongol roots. The garb is different too. Nearly all the women wear the *chuba*, a wrap-around dress of purple-brown broadcloth, covered by the usual figure-concealing apron. The men, in similar garments, have rolled down the tops and tied them at the waist, sometimes going naked above, and the children are in jumpers and quilted pants. A smoky odor rises from the crowd, which has pressed past the triage station and into the little courtyard beside the gompa. The shouts of children and the snap of prayer flags in the wind punctuate the murmurs of the throng.

Two policemen in stiff camo uniforms have climbed a wall to survey the goings-on, swagger sticks under their arms, disdain in their eyes. Lowland Nepalis, they give the impression of having frowned continuously since birth. A mob of choughs (pronounced "chufs") swirls overhead. They are little crows as black as tar. From long experience, they watch the crowd expectantly. Activity at the gompa usually signifies a religious ceremony, with offerings of yak butter and tsampa set out where the birds can eventually pillage them. They swoop low, cruising the throng, but detect no feast. Gradually they lose interest and coast away.

A drunk wanders in. He is aged, disheveled, and perhaps doddering as well as soused, and one of the grim policemen drives him off. When the old man staggers back, the cops look away, the better to preserve their dignity, and Jigme briefly leaves his post by the triage table to escort him beyond the courtyard. But soon the old man returns, tottering dangerously past tent stays and random chairs. A girl of perhaps six, no more than a yard in height, takes charge. She barks at him sharply and turns him around, pushing on his hips. She gives him a shove, scolding nonstop, and harries him off. This time he does not return.

At our morning circle Roshi had said to the clinicians, "Use this as a generative moment in your practice of medicine." She acknowledged that not much would be normal, especially for the Westerners. Diagnosis always involves uncertainty, but here the ambiguities of translation compound it, and culture compounds the ambiguity even more. If, say, "dizzy" is indeed a correct translation of what the patient said, what does "dizzy" connote in Dolpo—is it a specific loss of balance or a general feeling of malaise? Or is it something altogether different? And what if "dizzy" should have been translated as "nauseated" in the first place? Time can be puzzling too. When your patients—virtually all of them—say that the onset of their ailments was "years ago," does that truly mean years, or might it be months? Can everything be of such long duration?

You grope for answers. The tools at your disposal include a stethoscope and three blood-pressure cuffs shared among the practitioners, only one of which seems to be reliable. You cannot order an X-ray, a scan, or blood work. You interview the patient, aided by an interpreter.

You perform the physical exam, resorting to touch, feeling the pulse, absorbing the look of the eyes and ears, the throat and skin; you palpate the abdomen and percuss the chest. This is part of what you came for, a return to medical basics.

As the means of diagnosis are limited, so are the treatments. There are no return visits, no follow up. Your pharmacy has limited supplies, although at our meeting this morning Charlie McDonald urged that no one think in terms of shortage: "Treat what you need to treat. Don't hold back." And Dr. Sonam cautioned that in Chinese and Tibetan medicine the patient expects to take a handful of pills all at once, a practice that may suit the use of herbal compounds but becomes dangerous with Western meds. Think of an entire course of antibiotics concentrated into one big gulp. The result could be horrific. So, he said, short courses are best, but don't forget to emphasize finishing the course even when the patient feels better. With pain meds, Charlie suggested, tell your patients, take one or two pills on a bad day and save the others for the next bad day. Be really clear about dosage. And when in doubt, trust your instincts. Also, take a break when you need one. Roshi Halifax closed with her medical mantra: "care over cure," which implicitly admits that not much is going to get fixed, that you do what you can while you can, and you do it through making a connection: "warm hand to warm hand." She closed with another, oft-repeated admonition: "strong back, soft front." The phrase encapsulates the ethic of the expedition, and not just of its medical practice: everyone is on call, at all times, to deal with whatever develops. You try to keep yourself ready to handle the hard stuff, but not by shutting it out. You let it in, along with the people entangled in it. You face up to things with calm compassion, and you act. The meaning of the phrase sharpens when you think of its opposite: "strong front, soft back," which describes going through the world armored against criticism and connection, chronically scrambling to conceal vulnerabilities, putting on a show to shield the lack of substance behind one's bluster. We've seen leaders like that, adept at hoodwinking the gullible, but while such falsity may succeed in the domain of social media and retail politics, it is useless on the loneliest side of the Himalaya, amid the rockscapes of Dolpo.

14

Roshi

JOAN HALIFAX voyaged through the sixties and seventies like a counterculture Forrest Gump. She worked with the renowned ethno-musicologist Alan Lomax when the folk music revival, which Lomax helped to launch, hit its zenith. She hung out with Timothy Leary and Richard Alpert (long since known as Ram Dass), who became a close friend, when they were turning on, tuning in, and dropping out. She helped Joseph Campbell pull together his integrated view of world mythology. (Campbell may not be as well known today as when he was a best-selling author, but his *Hero with a Thousand Faces* left a stamp on popular culture by providing the essential plot for the first *Star Wars* movie.) Along the way, Roshi also did fieldwork among the Dogon people of Mali, the Huichol of Mexico, and prison populations in south Florida. Having earned a PhD in medical anthropology, she, along with Stanislav Grof, to whom she was briefly married, conducted clinical experiments in which they administered LSD to the dying. Their collaboration produced *The Human Encounter with Death*, which appeared in 1977, with an introduction by Elisabeth Kübler-Ross. A feature of the book was its exploration of the similarities between the psychedelic experiences of dying Westerners and accounts of death related in the Tibetan Book of the Dead and other, kindred wisdom traditions.

Roshi continued her work on death and dying in various guises. She became a hospice worker, a trainer of hospice workers, a collabora-tor in the development of new clinical approaches for end-of-life care, and a widely published counselor to caregivers, medical professionals, and many dying patients themselves. Her genius has been to help people face death squarely, without fear. When it was an unspoken presence, she spoke about it. When it cast its shadow, she turned on a metaphorical light and helped its imminent victims, their friends, and their families see it clearly. Her byword "care over cure" grows out of realism: when cure is out of reach, it is out of reach. No fairy tales.

Whether the prospect is for death in the near term or for long-running incapacitation, the care she advocates is directed toward making the present moment and the days that remain as good as they can be. This might mean rejecting the long-odds gamble of radical surgery or other interventions that require more tubes, machines, and drugs, and that induce a progressive loss of consciousness and connection. Such an approach may sound excessively dry-eyed and analytical, but the glue that holds it together is compassion. Roshi herself embodies the resulting paradox: grit and toughness married to warmth and empathy.

I have come to Dolpo partly to watch these medical values in action and partly to think about the ethics of hospice in relation to caring for the planet. The proposition is this: what if we focus less on "saving" Earth and more on caring for it? What would change in the work of Earthcare, and more particularly, what would change for the caregivers?

Two things bear repeating. First, Earth is not dying. The planet will go "cycling on," in Darwin's phrase, and living things will continue to evolve "in forms most beautiful and most wonderful" for eons to come. The entity that is dying is the "creation" of which we humans are a part, the most splendid and diverse proliferation of life-forms the planet has yet produced. The steady destruction of this bounty, the Sixth Great Extinction, is demonstrably true whether you believe in evolutionary science or a divine Creator. And not all life will die, just the wilder, more fragile, and, by most lights, more beautiful portion of it.

A second essential point is that hospice does not mean giving up. It means letting go of attachment to ultimate outcomes and making the best of the present. Interestingly, studies show that patients who go into hospice live at least as long and often longer than patients in comparable health who submit to drastic interventions. As Atul Gawande puts it in *Being Mortal*, "The lesson seems almost Zen: you live longer only when you stop trying to live longer." Meanwhile, the quality of life enjoyed by hospice patients, in their last weeks and months, is incomparably superior. And so is that of their loved ones and caregivers, a point that deserves emphasis. While hospice nurses do not work less hard or with less dedication than their counterparts in intensive care, they are three times less likely to suffer major depression. I take this to mean that they are less vulnerable, not just to black moods, but to burnout, which happens also to be an occupational hazard in the line of work I have pursued most of my life—conservation of the environment. While my colleagues and I may have won some battles, we acknowledge that the war has gone rather badly. I know I am not alone in feeling ground down by an unrelenting stream of reports about lost species, rising seas, bleached coral, melting ice, disappearing insects, and all the other contents of Pandora's environmental box. I have begun to wonder if by changing my focus from *winning* to *serving*, as a hospice worker might do, I might feel a kind of liberation. And I wonder if I might not see something similar among the clinicians of Halifax's Nomads Clinic.

Roshi possesses a nomad's restlessness. The early 1980s found her on a trail in far-eastern Nepal bound for the mountaineering basecamp at Kangchenjunga, the world's third-highest peak. With her was Daku Tenzing Norgay, wife of the man who, with Edmund Hillary, was the first to summit Everest in 1953. The two women became friends. After Tenzing Norgay died in 1986, Daku asked Halifax to carry his ashes on a trip that Daku, for want of a visa, was unable to make: the kora around Mount Kailash. She believed that taking Norgay's ashes in the form of a *tsa tsa*—a small molded sculpture—to Dolma Pass, the high point of the kora at an altitude of 18,471 feet, would help assure for him a favorable rebirth.

What began as a pilgrimage became for Halifax something like the hero's journey that her friend Joseph Campbell identified in mythologies from around the world: a trip to the edge of existence, a near disintegration, and a remaking of a wiser self. Halifax turned forty-five in the course of that journey, and noticing "that those few westerners who were on the trail were in their twenties," she began to question the wisdom of what she'd set out to do. In her telling, the greatest challenge was not the washed-out roads, mudslides, or hard walking. Nor was it the terrifying truck rides with men she presumed to be bandits. It was the food. And its lack. For four months she subsisted on tsampa, nettles, rancid yak butter, and salt. When she returned to the States, she weighed less than a hundred pounds. It was only later that she realized that the journey's gantlet of hazards and trials had welded her to the Tibetan world.

Fast-forward past long periods of study with the Zen masters Thich Nhat Hanh and Bernie Glassman to the present. Our trip into Dolpo is her seventeenth expedition in alliance with Tenzin, and before that, there were other journeys supported by other partnerships. She credits Peter Matthiessen, a long-time friend and fellow Zen practitioner, with steering her toward Dolpo. He would have come with us, she says, except he didn't last long enough—Matthiessen died in 2014. Viewed one way, astride her horse and led by Buddhi, the Sherpa guide who devotedly attends her year after year, she seems a recluse: a human speck on a narrow trail in the oceanic barrens of the Himalaya. But that is not her only self.

A recent count showed Roshi with more than twenty-five thousand followers on Facebook and forty thousand on Twitter, perhaps fewer than a rock star but more than many politicians. Except when she is in retreat or in the wilds of Nepal, she posts on several social media platforms every day, and she emails at all hours. It is no accident that her long-time aide, Noah Rossetter, a veteran Nomad, is an IT wiz, well-versed in the ways of the internet. Her networks, both electronic and relational, are formidable. They include patrons with such names as Rockefeller and Kluge. Her legions of admiring "friends" give her clout: following the 2015 Nepal earthquake she raised hundreds of thousands of dollars for relief efforts in a matter of days. Plus she has built a meditation and retreat center in Santa Fe that is like a small college in the abundance of talks and courses it offers. Yet she wears these accomplishments lightly, as I learned a decade ago when my nephew and his fiancée asked me to officiate at their wedding. I needed a script for the ceremony and so I asked Roshi if she had one I could borrow. She said, "Sure. Use this. Just take out all the Buddhist shit, and you'll be fine."

Most people like to take shortcuts when they can, but Roshi seems disinclined toward the path of least resistance, opting usually for the longer reach, the loftier goal. If more resources are required, they can and will be found; they might need a little more looking for, more calls and emails, more connections made, more look-'em-in-the-eye, not pleading but upfront telling "you need to do this," in so many unambiguous words.

Let it be said that she has detractors. She has left a wake, and a number of former friends now stand at a distance. But her accomplishments remain impressive, and more than most of us, she has relieved suffering in many places, while also inspiring a small army of students and colleagues to relieve more. I have known no friend more generous. Or more resilient. A decade ago, in a freak fall in a colleague's bathroom, she shattered her femur and trochanter, the knob of leg bone that fits into the pelvis. Surgeons replaced a length of unmendable bone with enough titanium to set off airport scanners. After the accident, I expected that the radius of her activities would shrink radically. She thought so too. And then the limitations, including loss of contact with the Himalaya, began to grate. She fought back, walking miles around the little mountain retreat that she maintains an hour and a half outside Santa Fe, close to where I live.

Eventually the question of other means of travel arose, horses in particular. A celebrity member of the Upaya community invited her to go riding at the celebrity's nearby ranch. The excursion amounted to a graduation exam for Roshi's reconstructed leg. She went, rode, and survived. The celebrity made her a gift of the soft and accommodating saddle she used that day. Since then, the saddle, renamed the "Jane Fonda BarcaLounger," has been essential equipment for every Nomads expedition. Her friends wonder—and Roshi does too—how much longer she can continue to meet the challenges of frozen mornings and long days on the trail, to say nothing of her pre-Covid, jet-fueled itinerary of lectures in Japan, workshops in Europe, and consultations in New York and San Francisco. Roshi likes to think that she will pass the mantle of Nomads leadership to Charlie McDonald, Wendy Lau, and others, but no one expects to fill her shoes.

As a teacher, she is known for timely prodding. Her nudgings may not always hit the mark, but she bats a high average. Often, when we assemble in morning circle, she throws out an idea or problem—not quite a koan—for us to chew on through the day's journey. One morning she said, "Keep this phrase in your mind and see where it takes you: 'Do not find fault with the present.'" The group mulled this for a moment.

Then she added, "I mean this especially for you, deBuys."

And so for the next half-dozen miles I kept returning to that idea: "Do not find fault with the present." I readily admit admiration for anyone who can, in good conscience, hold to such a view. It requires an abundance of detachment. Also equanimity. But all that day and frequently thereafter, I doubted my ability or even my desire to attain that state, the faults of the present being so manifest, the need for response so clear.

15

Hospice for a Mouse

MIDWAY THROUGH the afternoon of the Tokyu clinic, the ever-smiling Thupten asked Roshi to tea, and she, seeing I was unoccupied, invited me to come along.

Thupten is our host in Tokyu. We are camped in his barley field. Our clinic occupies the courtyard between his house and the gompa, which his family has looked after since the first temple was erected there, as the family reckons, six hundred years ago. The present building, the last of a succession of structures, is already a century old.

As with snow-country farmhouses throughout the world, Thupten's living quarters comprise a second story, set above dark stalls where livestock can be quartered for winter warmth. We climb steep stairs to an entry room where we shed our shoes and outer garments. Then Thupten shows us into his shrine room. Sunlight pours through large, south-facing windows to illuminate a large brass gong and an array of cymbals and drums, one of which, I learn, is made from two human crania joined crown to crown and covered at their bases with taut skins. This instrument, the *damaru*, is held in the lap, while the drummer strikes the skins with either hand. The ceiling is split-cedar on squared wooden beams, a handsome achievement in an area that grows no sawtimber. The floor gleams as though made of varnished teak or mahogany, and it is some minutes before I recognize the faux-parquet linoleum for what it is. The wall at the head of the room tricks the senses too. It is painted to replicate the altar wall of a gompa, with a grid of cubbyholes flanking a central display of sacred images. Sacred texts, or *pechas*, and their associated commentaries fill the cubbies. Not until Thupten's graceful wife, shy and beautiful, brings a pot of tea and Thupten pours our cups, is the secret of the room revealed.

Roshi asks Thupten about the pechas in his possession. He goes to the painted wall and, as though by magic, swings it open. The cubbyholes are *trompe-l'œil* illusions, painted on a flat surface. Thupten has opened a cabinet, one of several in the wall, and from it withdraws two documents, one a modern, printed text, the other a heavy, atlas-size compendium that he strains to lift down. Its pages are parchment, and may represent the skins of an entire herd of goats. The unbound book is inches thick and badly abraded at the back, perhaps gnawed by rodents. It is hand-lettered in small, dense script of a style that Thupten says has not been used for centuries.

The book was smuggled out of Tibet when the Chinese seized the country in the 1950s. Had the invaders found the book, they would have burned it, as they burned literal tons of Tibet's ancient treasures, a pillage comparable to burning the libraries of the great medieval abbeys of Europe. Here we held a treasure in our laps, a symbol both of the antiquity of Tibetan traditions and of their endangerment. Under its "One China" policy, the Chinese government has suppressed Tibetan cultural identity and the separatist impulses it generates. Tibet's monasteries have been drained not only of their treasures but of their monks and, more generally, their importance. A monastery that once housed a hundred monks might now have three, and so the chants, *pujas* (ceremonies), and dharma talks that formerly connected the monastery to the surrounding community have in many cases dwindled to irrelevance. Chinese TV, internet, and roads, meanwhile, have spread to the farthest corners of the plateau, inculcating the young with the values of the modern Chinese state. Taken together, these changes mean that the villages of Upper Dolpo constitute a final redoubt for traditional Tibetan culture. Insulated by distance and terrain from both Beijing and Kathmandu, they remain more Tibetan than Tibet.

We return a second time to Thupten's shrine room the next morning for meditation and a puja that consists of chants performed by Thupten, his amiable cousin, Lama Ngawong, and a third man with soulful, drooping eyes whose name I do not learn. Lama Ngawong has traveled with the expedition from the outset. Most of the year, he presides over a Buddhist temple in Kuala Lumpur. He seized upon the opportunity afforded by our journey, not only to return to Tokyu, where he was born, but to visit Shey, our ultimate destination, and a gompa not far from Shey called Tsakang, where, barely a teenager, he underwent the three-year retreat through which he attained the status of lama. Many Buddhist clerics are educated and trained in Nepal yet earn their keep in distant corners of the world. Not all of them provide assistance to their home communities, but the remittances from those who do, like

Lama Ngawong, can mean the difference between survival and abandonment for gompas and homesteads alike.

We fill the room, sitting on pillows and pads, crammed together. The chants begin. The first, said to have been handed down from Guru Rinpoche, who brought the dharma to Tibet, builds like a fugue through waves of complex repetition. The singers glide through its strange musical slaloms, evoking a time and a world long past. The second chant is dronelike, in two-beat meter. And the third, after a slow, languid beginning, suddenly explodes into an alarming cacophony of percussion: gong, bell, cymbal, rattle, and drums, including the damaru, every instrument at top volume. After the outburst, the chant resumes at a faster pace, until a second pandemonium erupts, and then a third, and a fourth, after each of which the tempo further quickens until the singers utter their syllables as fast as tongue and breath allow, their words becoming a torrent. And then abruptly they stop. In the following silence, we hear the creaks of the building and the wind outside the walls. Collectively we exhale, as tension drains away.

We come again to Thupten's house that evening, our last in Tokyu, for a performance of traditional music and dance. We gather, not in the shrine room, but in the large common room that combines kitchen, living, and sleeping space. Westerners and Nepalis alike, we fill the benches lining three walls of the room, as well as most of the floor space around the low-slung, stamped-metal Chinese stove that one of Thupten's daughters periodically stokes with yak dung to produce a wan heat. Several fluorescent tubes, powered by solar-charged batteries, provide light.

Opposite us stands Thupten's radiant wife and four other women in the traditional garb we have grown used to seeing, but now the outfits they wear are brightly clean, the colors brilliant. A stout roof post stands between the women and Thupten, Ngawong, and their companion from the morning chant. Thupten holds a traditional Tibetan lute, or *dranyan—dran* for sound, *yan* for melody. Between a mandolin and a guitar in size, it has five strings anchored to wooden tuning pegs. Above the pegs, the carved headstock of the instrument curves forward in a representation of a horse head endowed with a luxuriant mane.

Each song starts slowly, with Thupten or sometimes Ngawong plucking the dranyan, the women singing in birdlike voices and dancing in intricate, mincing steps. The men, more animated, belt out lyrics as the song quickens in tempo, faster and louder, until the line-dancing women gasp for breath and the men almost shout. When each song reaches its apogee of frenzy, the master of the dranyan, be it Thupten or Ngawong, ecstatically flails at the strings like a Western rock and roller. I never learned the meaning of the lyrics but imagine they might have been carried from the arid heart of Mongolia, across the Tibetan Plateau, and down into the rockbound valleys of Dolpo. I tell myself it is the music of people accustomed to gathering in lonesome canyons under an infinity of stars.

The strangest punctuation of our evening's concert is the appearance of a dun-colored, long-tailed mouse. The creature enters the room from an unlit passageway that may have led to a pantry. From dark to light it comes, from silence to loud noise, and from the safety of its hiding place into a room crowded with humans. A ripple moves through the people on the floor as they scoot aside to make way for the mouse, which seems intent on its destination. It darts on quick feet through the maze of bodies, pillows, and rolled-up jackets until at last it reaches the central stove. There, not so close that it would overheat but near enough to feel a steady warmth, the mouse sniffs the air and surveys its surroundings. It then lies down and is still. Even as the music plays on, we stare incredulously at the bold mouse, murmuring our amazement and sharing quizzical smiles. The mouse sleeps.

One cannot help but invent a story to explain so strange an event. I tell myself that Thupten's household must be so peaceful that even the mice are unafraid. The lowliest creature is as privileged as any person to enjoy a place by the fire. Only after the evening's entertainment has concluded and we file back into the night am I disabused of my narrative. Word passes that the mouse was not sleeping. It was dead. Inexplicably, it had come from its lair, traversed a roomful of people, and lain down beside the fire, where it breathed its last. We had no idea what we were seeing.

16

Lumber Yaks

DAYS AGO, when we were setting camp in the vale between Bogu La and Numa La, the gods of slapstick sent us a joke.

An early task in a new camp is the pitching of a latrine tent. Amon, an acupuncturist from Germany, was among the strongest walkers in our group and consistently arrived in camp before the rest of us. On this day, he elected to use the latrine. Once inside, he zipped up the tent flap behind him.

Then came the yaks loaded with lumber.

They topped a rise above the camp and shuffled down through our clutter of piled gear and half-pitched tents, an irregular line of at least twenty woolly, red-eyed beasts, each burdened with massive, fresh-sawn pine timbers. Loads varied. Most of the yaks bore a pair of six-by-eight-inch, eight-foot beams, one on either side, the kind of beam that spans a wide doorway or bears an upstairs floor. Others carried bundles of two-inch planks, or combinations of planks and beams. With few anchor points, the loads waggled atop the animals, sometimes dragging a rear corner on the ground. All the wood was moist to the touch, green and very heavy, yet the animals strode indifferently onward in their semi-waddling downhill gait, which set the lumber to waggling even more.

The progress of the caravan boiled with random motion, each yak navigating its own path, their trajectories unpredictable. The ends of the beams danced like pistols held by drunks. Several yaks veered toward the latrine tent, and Amon must have sensed their coming, for when an errant beam snagged a tent stay, the door was already unzipped and Amon came leaping through it, hopping two-footed and hitching up his pants, even as the tent came down. He landed in the path of another yak, which shied into a pile of duffels. The herdsmen, following afoot and leading their horses, joined the general laughter as they whistled their charges through the camp. Amon laughed loudest of all.

It turned out that one of the herdsmen was Lama Ngawong's

younger brother, whom I happened to meet later in Tokyu, where he and his companions paused for a day to rest their yaks. He told me they obtained their lumber close to Phoksundo and were headed north along the same route that we would soon follow. In four days' time they would reach a Chinese border post, and there, amid a sprawl of tents stocked with the commerce of the far Himalaya and under the unsmiling gaze of Chinese border guards, they would rendezvous with other traders coming by prearrangement from inner Tibet. They would exchange their lumber, along with a quantity of rice, for cash, sodas, beer, and other miscellany. Such trade, albeit for different goods, has existed between Dolpo and Tibet for centuries, salt from the plateau being exchanged for barley and other grains produced in Dolpo and other, more distant districts of Nepal. The trade being regular, the relationships between partners persisted for generation after generation so that families at either end came to view their trading partners as virtual kinfolk, linked in an ancient bond they called *netsang*. The trails we have been walking are salt trails, and the network of netsang extended not just from Tibet into Dolpo but beyond Dolpo into Nepal's lowlands, where rice and other goods were added to the web of interaction.

In Dolpo, farming alone cannot sustain a family. The land is too austere. Trade was one way of supplementing a family's livelihood. Livestock grazing was another—and the best winter rangelands were always in Tibet. Some say that the need for economic diversification explains the practice of polyandry, formerly common in the region, by which a woman takes multiple husbands, all brothers, perhaps one to farm, one to herd, and one to trade, the last two being absent for long periods.

The old ties of netsang were much weakened when China closed the borders of Tibet, restricting commerce to a frantic week or two per year. The closed door of the border also barred Dolpo's yak herders, many of them nomads, from the Tibetan pastures on which they depended, thereby guaranteeing more intensive—and excessive—use of Dolpo's wild rangelands.

Nowadays pine timber is a principal commodity on the trails to China. At the border Lama Ngawong's brother will unload the lumber from his yaks and pile it on the trucks of his partners, and it will be carried off into the treeless plateau. Never mind that the timber was cut within the boundaries of Shey-Phoksundo National Park, where logging, in theory, is permitted only to provide building timbers for schools, hospitals, and other community projects. Never mind that, if these timbers were cut under the authority of a permit, it was a false one. The park and its regulations exist on paper, not so much on the ground. Such is the reality of "protected areas" in much of the world. In weeks to come, we will see dozens more caravans of Phoksundo timber, their scores of yaks bearing away the best fruit of Dolpo's forests along the old salt trails to China.

17

Aunts and Uncles

THE SPIRIT of the trail changed for me, more than a decade ago, when I received permission to use a cabin, usually reserved for field ecologists and trail crews, in the backcountry of a national park. I wanted to investigate an archaeological site, to determine if a certain corridor of arranged stones aligned with the dawn rays of the winter solstice. The overnight lows that December hovered near zero on the Fahrenheit scale, but a fire in the wood stove made the cabin cozy. Next morning, the angle of first light on the ancient shrine confirmed that it functioned, at least in part, as a solstice observatory. But that's not why I tell this story.

The outhouse that served the cabin had splendid graffiti, featuring quotations from the likes of Rachel Carson and Aldo Leopold. Most prominent were the opening lines of Walt Whitman's "Song of the Open Road":

Afoot and light-hearted I take to the open road,
Healthy, free, the world before me,
The long brown path before me leading wherever I choose.

His lines distill the spirit of the trail into verse, and they were in my mind as we set forth from Tokyu, our train of walkers and pack animals stringing out for a mile or more as we climbed toward yet another high pass, this one called Chhoila.

I thought back to the attribution scrawled beneath the quotation on the outhouse wall. It said, simply, "Uncle Walt." There was no need to add "Whitman."

Strange, I thought. I too called Whitman "Uncle Walt." So did many of my friends. Together with the backcountry graffiti writer we thought of him in avuncular terms. Admirers of his poetry might say that it is hard not to. Uncle Walt speaks to us as though we were kin:

I am with you, you men and women of a generation, or ever so
    many generations hence,
Just as you feel when you look on the river and sky, so I felt,
Just as any of you is one of a living crowd, I was one of a crowd,
Just as you are refresh'd by the gladness of the river and the
    bright flow, I was refresh'd

His voice is the voice of an old friend. Sometimes at the writing table I hear his cadences, his compendious lists, his easy informality and American diction.

Musing about the graffiti in the backcountry outhouse led me to think about other aunts and uncles. Eventually, I wrote down their names. I listed people living and dead from whom I got something essential, something without which I would not be who I am, or at least who I think I am. Many of the names were private people, significant only to me. Many were historical figures known to everyone. They stood for something I valued, held a place in my pantheon, helped me sort out the world and make a path in it. The list grew to three-hundred-sixty-odd names before I stopped, but I could have gone on. All those shoulders we stand on.

Alphabetizing the list was amusing. Duane Allman came after Dante Alighieri. Fyodor Dostoevsky preceded Frederick Douglass. I began to imagine random five-name sequences as line-ups for basketball teams: Marvin Gaye, the anthropologist Clifford Geertz, Mohandas Gandhi, Theodor Geisel (Dr. Seuss), and Geronimo. Who would play the low post? Who would shoot threes? It got silly.

But it was also instructive, and I commend the process. Sometimes, especially walking in a big landscape with no one near, I begin to feel that I am not so much an individual as a throng. I sense the aunts and uncles around me, a posse of ghosts. We troop down the trail in a little clot.

Such were my feelings on the long descent from Chhoila Pass (16,600 feet), which we topped on our first day out of Tokyu. Our path led toward a north-flowing river that we would follow to the village of Tinje, hard by the Tibet-China border and the site of our next clinic. From the top of the pass the vistas were lunar: naked mountains, all gravel and rock, stretching to a gauzy horizon. No tree stood up from the land for scores of miles. There was hardly a bush. Ragged mats of tundra clung to the gullied slopes. I heard the hiss of the wind and the crunch of my boots. The world was skeletal, bone without flesh.

In the thin high air the light seemed to change. I watched a lammergeier, the immense, bone-eating vulture of heights and wild places, soar against a sky that grew deeper and bluer. Shadows endowed the bleached vistas with a rich, purpling palette. Shorn of all concealment, the land felt alive, as though its slow tectonic rising and slower wearing down might be directly sensed. I saw myself as flotsam on a sea of crustal upthrust, and I thought about the omission of Alfred Wegener from my original list of aunts and uncles. Back when I drew it up, I had not appreciated how profoundly geology shaped not just the actual world but my small, individual view of it. I realize now that, even before I knew the phrase "plate tectonics," a sense of the living land affected how I looked at a map or a mountain. It underlay my faint understanding of the course of rivers, which so often follow geologic faults or embody the sawlike abrasion of water as the land slowly rises. The effects could be seen wherever one's eye might rest, especially in the Himalaya, where the land is young and ever-changing. I resolved to add Wegener to my roster of aunts and uncles, and I made a note also to add Harry Hammond Hess and Tuzo Wilson, among others, for they were among the pioneers whose work recast continental drift as plate tectonics and made possible the modern rediscovery of Earth.

18

Hidden Figure

SEAFLOORS ARE young. Continents, for the most part, are old. This is a revolutionary insight, divined in the 1950s and '60s, when, as schoolboys, my classmates and I were learning to "duck and cover." On random mornings an air-raid siren would wail, and our teacher would quickly tell us to huddle under our desks. Surprisingly, "duck and cover" and the news about the seafloor were connected.

It was the era of the Cold War, and a nuclear contest between the United States and the Soviet Union seemed probable, if not imminent. At school we learned to crawl under our desks as the best defense against being blown to bits by a nuclear bomb. The exercise was futile and absurd, but we children did not know that.

There was more. Down at the end of the road where I lived, giant piles of dirt began to appear. It turned out our neighbor was excavating a fallout shelter. He was a psychiatrist. My dad thought he was crazy. My dad also thought psychiatry was crazy. In fact, we were all crazy. The strategic military doctrine of the United States and the Soviet Union was called MAD, or "Mutually Assured Destruction," which meant that, even if you blow me up, I can still blow you up, and I will. In a world gone MAD, the few defenses included "duck and cover" and fallout shelters. If you think the world was insane then, you have to agree it remains so, for MAD still underlies nuclear deterrence.

One of the most potent weapons of the early Cold War was the missile-toting submarine. Its ability to hide in the ocean depths made it harder to destroy than land-based missiles or fleets of airplanes, which were the other two "legs" of the so-called nuclear triad. In 1946, when the smoke of World War II had barely cleared, the US defense establishment created the Office of Naval Research (ONR) to better understand the undersea environment as a theater of war. A first step was to map the seafloor. The logic was simple: by locating bumps and crannies in the floor of the ocean, we would improve our ability to conceal our submarines, as well as to find those of our enemies.

Hundreds of scientists rallied to the work, which was deemed as cutting edge as aeronautics, another big recipient of military largesse. Private institutions led the charge. Before long, such laboratories as Woods Hole on Cape Cod, Lamont at Columbia, and Scripps in California were flush with ONR funding, which the National Science Foundation, founded in 1950, soon augmented. Research ships sailed tens of thousands of kilometers pinging the ocean bottom with sonar, towing magnetometers, and setting off explosions to record reflected shockwaves on seismographs. Military and scientific interests intertwined. Maps of magnetic anomalies, which produced surprising patterns for researchers to interpret, also suggested new ways to find big submerged chunks of metal such as submarines. Nuclear testing, including the need to detect explosions far away in the Soviet Union, drove major advances in seismology. By the mid-1960s, the US had installed hundreds of seismic stations in the deep sea, so that neither the Soviets, nor the Chinese, nor anyone else could keep a thumping-big explosion secret. As a byproduct, the network gave new insight into the location and frequency of suboceanic earthquakes and the composition of the seafloor (because sediments and bedrock reflect shockwaves differently).

The data flooded in. Before long a new generation of marine scientists had mapped the seafloors with a degree of geophysical detail that their land-bound colleagues envied. Oceanographers joked about the "familiar ocean basins" versus "the mysterious continents."

Familiar the oceans may have become, but their basins turned out to be far different from what cartographers expected. The seafloors were thin, much thinner than the continents. They were geologically young. They were topographically rough but regular, often oddly symmetrical. It emerged that the Mid-Atlantic Ridge, which had been known since the 1870s, was not unique. Other oceans had similar submerged "mountain ranges," and a depression resembling the rift valleys of the continents cleaved down the center of each one. These valleys were measurably hotter than their surroundings. Magma presumably lay close beneath them. Perhaps strangest of all, the one hundred million depth soundings that oceanographers had measured revealed ocean trenches of fantastic profundity, and the "deep focus" earthquakes that seismologists were recording seemed to center in and around those trenches.

Blind men describing an elephant had it easy compared to scientists trying to assimilate the new learning in a coherent picture.

The huge haul of data from research ships overwhelmed the institutions and individuals who tried to process it. Across many subdisciplines—topography, gravimetry (the measurement of variations in gravitational force), magnetometry, heat-flow analysis, seismology, and more—the data could scarcely be stored before the next voyage set out. What it all meant would have to be sorted out later, perhaps much later. Making connections across different domains of knowledge and lines of inquiry would come after that. The task of relating the new knowledge to stale, old theories about Earth history such as drift, contractionism, and permanentism languished near the bottom of geology's to-do list.

Stories get told for two reasons: because they are important and because they fit how we like to see the world. The story of Darwin's discovery of natural selection has both qualities. Not only does it concern one of the most powerful theories in science, it also has a noble and conflicted protagonist. Darwin departs society to undertake a daring, marvel-filled journey into the unknown, and then returns to share his hard-won wisdom with the rest of us. He even has a rival, Alfred Russel Wallace, whose parallel discoveries briefly cast Darwin's ultimate victory in doubt.[3] As Joseph Campbell might point out, Darwin's saga has the structure of myth, the most powerful kind of story there is.

The theory of plate tectonics, like that of natural selection, is one of the greatest achievements in science, yet its story is seldom told because the narrative is difficult. For starters, its elements are hard to understand: magnetic anomalies will never be as vivid as the finches and iguanas of the Galapagos. But more important is its lack of a single, valiant hero. Alfred Wegener makes a good character, but he was an unheeded prophet who failed to capture the final insights. The ultimate achievement of plate tectonics was corporate—not in the sense of big business, but of a collective endeavor involving many people, which is how most science gets done. In the story of plate tectonics, multiple names stand out, including those of H. H. Hess and J. Tuzo Wilson. More on them, and some of their colleagues, later. An additional name,

---

3 In the late 1850s, Alfred Russel Wallace (1823–1913) independently reached conclusions similar to Darwin's concerning the evolution of species. Their rivalry could have produced acrimony and bitterness, but Wallace soon acknowledged Darwin's preeminence and his grace in doing so set a standard for professional gallantry that has rarely been equaled.

a woman's, should also appear in accounts of the theory's development, but often doesn't. It belongs to Marie Tharp.

Her father was an agronomist who mapped soils across the American heartland, and Marie Tharp became a mapmaker too. In 1943 she entered a master's program in geology, which opened to women only because military service had depleted the available pool of men. A sympathetic professor, knowing her gender would bar her from field-work, urged her to take up drafting. Years later, having landed a job at Columbia University's Lamont Geological Observatory (now Lamont-Doherty Earth Observatory), Tharp was charged with mapping data that streamed in from ocean-borne research expeditions. Her boss and principal collaborator, Bruce Heezen, was often absent on such voyages, which were entirely male undertakings, women then being banned from virtually all research vessels.

In 1952 Tharp began plotting soundings from six transects across the North Atlantic. Although oceanographers had long been aware of the Mid-Atlantic Ridge, no one had mapped it in three dimensions, fully capturing its topography and creating a visual image of its mountains and valleys. When Tharp did this, she immediately recognized a starkly V-shaped feature running north and south, slicing down the crest of the ridge. She told Heezen it looked like a rift valley.

For Heezen, the implications were intellectually and reputationally toxic. A rift implied a place where opposing landmasses were pulling apart from each other. It was the kind of nonsense posited by advocates of continental drift. Lamont in those days was headed by Maurice Ewing, an adamant "fixist" for whom the stability of seafloors—and of the rest of Earth—was dogma. He had made Lamont a citadel of resistance to drift, and Heezen needed no coaching to recognize that countenancing Tharp's hypothesis was a kind of heresy. He conde-scendingly dismissed Tharp's idea as "girl talk."

But he could not banish the facts. When another researcher's map of earthquake epicenters was superimposed on Tharp's map, the align-ment of quakes with Tharp's putative rift was unmistakable. Tharp soon demonstrated that the rift continued into the South Atlantic, and Heezen was obliged to come around to her view. They published their first physiographic map of the North Atlantic in 1957.

Tharp extended her mapmaking further. She showed that the Mid-Atlantic Ridge extended for ten thousand miles through the North and South Atlantic Oceans. Ultimately it would be found to connect to ridges and rifts in other oceans extending for four times that length, girdling the planet like seams on a baseball. The rifts and ridges turned out to be Earth's largest physiographic feature. No one had previously known it existed, and no one had "seen" it until Marie Tharp drew it on a map. But what did it ultimately mean? For a time, neither Marie Tharp nor anyone else could say.

19

Geopoetry

WITH THE outbreak of World War II, Professor Harry Hammond Hess departed Princeton University for service in the US Navy. During much of the war he commanded the attack transport *Cape Johnson*, delivering troops and supplies to multiple battle sites in the island-hopping campaigns of the South Pacific. From the vantage of the *Cape Johnson*'s bridge, Hess used his "echo sounder," an early form of sonar, rather more than was necessary. Occasionally, as he crisscrossed thousands of miles of ocean, he altered his course, deviating from the expected route to record ocean depths in areas of particular interest. His wartime data gathering led to the discovery of submerged, flat-topped volcanoes, which he called "guyots," after Guyot Hall, the flat-topped home of the geology department at Princeton.

Hess returned to Princeton after the war and continued to teach while also serving as a committeeman on high-level national defense and science policy. More than perhaps any of his peers, he strove to integrate into a global model the new knowledge about ocean trenches and ridges, including their conspicuous rifts. By 1960 his concept was complete enough for him to present it to the world. But he didn't call it science. He called it "geopoetry," to disarm his critics.

Hess's fellow geologists eventually referred to his conception as "seafloor spreading." The main elements were these: Magma wells up from the interior of Earth at the mid-oceanic ridges where it hardens and hydrates, forming the basaltic material of the seafloors. The production of seafloor is continuous, driven by magmatic convection in the mantle of Earth, causing seafloor to spread symmetrically outward from the ridge. As the ridge material cools, it contracts and also erodes. As a result, the farther new seafloor gets from its point of origin, the more it subsides toward the level of the surrounding "abyssal" plain. The slow conveyor-belt movement of seafloor away from the ridges eventually reaches trenches where the material of the seafloor is drawn downward—"subducted"—and recycled into magma.

Hess's geopoetry explained why seafloors were geologically young—a hundred million years or so were sufficient to complete the journey from birth in a ridge to death in a trench. Hess also removed one of the weightiest objections to drift. Neither Wegener nor Holmes (who first offered magmatic convection as a driving force for continental movement) had been able to explain how continents plowed through Earth's crust. Hess's hypothesis was completely original. The continents did not plow through the crust, he said; they were less dense than the crust and they floated on top of it. It was the crust that moved.

John Tuzo Wilson, a Canadian who had studied with Hess at Princeton in the 1930s, lent support to his mentor's ideas. Wilson was a man of energy who flew up stairs several steps at a time. His mind worked that way too, making leaps. He expanded on Hess's geopoetry with vital insights, one of which came to him on a trip to Hawaii. The Big Island, with its active volcanoes, is geologically the youngest part of the archipelago. Progressing north and west along the chain, each island is older than the one before it and more eroded and worn down, until at last the islands aren't even islands, only seamounts that fail to top the waves. Wilson reasoned that a "hot spot," a magma column, underlay the seafloor where the Hawaiian islands had formed, and that while the column stayed fixed in place, periodically erupting, the spreading seafloor moved over it, gradually carrying away the islands that the eruptions created.

If this was correct for the Hawaiian archipelago, it should also apply to other mid-ocean island chains: the oldest islands should lie farthest from the mid-ocean ridges where their respective seafloors were generated, and each island should be sequentially older as distance from the ridge increased. Wilson examined the globe's volcanic islands—the Faroes, Azores, Cape Verdes, etc.—and showed that his hypothesis held true in virtually every case. Thanks to Wilson, seafloor spreading began to leave the realm of "poetry" and acquire the heft of fact.

Hess and Wilson were rediscovering Earth. Their ideas depicted a planet different from any that humans had previously conceived. Proof of their formulations remained wanting, but contributory evidence was piling up. When the proof came, it would verify that the planet's youth and vigor were best expressed in the depths of its oceans and that the continents, where Sapiens dwelled, preserved Earth's most geriatric features. An exception, of course, is the Himalaya, which is as dynamic as any ocean ridge.

20

The Road to Tinje

PREM, AS Roshi likes to point out, looks like Attila the Hun. That he is a small Attila has not diminished his confidence. He carries himself as though he were the match of anyone, and if attitude can conquer size, he may be. Tenzin, his nominal boss, outfits the expedition. He provides the gear, handles the arrangements, obtains the permissions, and—this is conjecture on my part—pays the occasional baksheesh that allows us to proceed. Prem keeps the field operation humming, supervising the muleteers, kitchen staff, and Sherpas, as well as the frequently clueless Westerners, myself included. He gives directions in a stream of orders and suggestions, never raising his voice because he never has to. People listen to him.

Prem briefs us on the next day's route and destination during a nightly ritual. At the end of dinner Prem enters the mess tent and all chatter stops, replaced by calls of "Prem Time! Prem Time!" He reports on the day's events (perhaps including how many patients we saw) and forecasts the day ahead. The forecasts include estimates of distance to be covered and altitude to be gained or lost, ostensibly so that we can prepare correctly. But the estimates are consistently, often shockingly, low, perhaps so as not to discourage us. Michael Lobatz says, "Prem Time is when Prem comes into the tent and lies to us." More than once, sixteen kilometers have turned into sixteen miles, an increase of nearly ten kilometers. And a day advertised as "more or less Nepali flat," while beginning and ending at roughly the same altitude, can require twenty-five hundred feet of hard climbing and as much descent.

Our next destination, Prem explains, is the remote community of Tinje, the largest of a cluster of four villages that are home to more than a thousand people. Six centuries ago three brothers from Tibet settled there. Prem adds, "Nepal royal family was never king in Tinje. Too far from Kathmandu. Nobody care. Dalai Lama always king of Tinje. Still is."

But other kings vie for control.

Imagine the dilemma of the foreign minister of Nepal. Besides serving a feckless government mired in corruption and bureaucratic delay, the minister must propitiate two giant powers, India and China, which enclose Nepal as an oyster encloses a pearl. Both nations desire Nepal's fealty and its resources. And they are jealous rivals. To please one is to anger the other.

Part way between Chhoila Pass and Tinje, we stop at a cluster of ruins where a bridge, long washed away, formerly crossed the river we have followed for two days. The collapsed structures of a trading post are spread over several acres. Oddly, the high walls of a large stone livestock enclosure look freshly straightened, their gaps repaired. We peer within and are astonished to see scores of large steel drums, neatly ranged. Even upwind in a steady breeze, the place reeks of diesel. The former livestock pen is now a fuel depot. But no vehicle or machinery is anywhere in sight. We've been two weeks on the trail and have not seen an airplane in the sky or encountered so much as a motor scooter on the ground. The only internal combustion has been in our cells. Nevertheless, we've heard rumors that the Chinese, eager to prospect Dolpo for rare earths and precious metals, are building a road from the international border down to Tinje. Or from Tinje to Tokyu. Or that the Nepali government is building one or both roads to counter Chinese influence. Neither Prem nor Tenzin seems to know which, if any, of these stories is true, but they tell us that, for sure, a new road is in the works that will connect upper Mustang to Tibet. It is already under construction, and its path from Nepal into the plateau passes only a few score miles to the east, as the lammergeier flies. So a cache of diesel fuel in the environs of Tinje makes an ominous kind of sense. The proximity of China has lately become palpable for all of us, thanks to the increase of litter beside the trail, consisting of Chinese food wrappers and soda and beer cans. And now it seems further underscored by thousands of gallons of fuel stashed behind stone walls.

We are still drowsy from lunch and not yet booted for the day's last miles when there appears a clamorous answer to the mystery of the depot. First comes a distant metallic rattle, then the hammering throb of an engine. The ground shudders. A giant yellow excavator suddenly crests a rise in the valley floor. The name Hyundai is blazoned on the boom of its bucket. Its caterpillar tracks clatter like gunfire, and its lurching, folded boom and claw-toothed bucket give it the appearance of a nightmare insect. The huge bucket brims with several fuel drums. Still more drums are strapped behind the cab atop the engine compartment. The apparition startles the mules, which bray and rear. The machine stops opposite us, its din subsiding, as though it—and not its invisible operator—wishes to regard us. Some of us begin to approach it, but then the engine pounds faster, the tracks resume their clatter, and the machine clanks away toward the depot.

We set out quickly for Tinje, our spirits jarred. Our path follows, in reverse direction, the deeply gouged tracks of the excavator. Soon the tracks give way, not to a trail, but to a broad scraping of raw earth that seems the first draft of what might one day become a road. No culverts are in place, and the creeks and gullies that drained across the scrape have either backed up into stagnant pools or carved gullies through mounds of loose dirt, according to their strength. Our walk becomes a long, glum slog.

The road is not our only worry. Charlie has mentioned that he might have to order his own evacuation. A suspected bone spur on his ankle has caused him intense pain. As a result, he has mostly ridden a horse these last days, but the custom in the Himalaya is only to ride *up* the steeps, not down them, a practice that avoids accidents and saves the legs of the horses. Even Roshi, with the obsessive assistance of Buddhi, descends the steepest trails on foot. Charlie has gamely limped through cumulative miles and in camp gimps about, but he is reaching his limit. Amchi Lhundup has slathered a poultice on his ankle. Our acupuncturists have riddled it with needles. Other doctors have poked and prodded. Everything helps, says Charlie generously, but nothing fixes.

Several miles from Tinje we notice groups of women on multiple paths, mere dots on the naked canyonsides. Their trails converge northward, toward the village. Through binoculars we see that each woman stoops under a bundle nearly as big as a fuel drum. The load rides on her shoulders, steadied by a tumpline to the forehead. The bundles consist of sticks, thin ones, from what looks to be a shrub. Evidently the shrub grows far from the village, and next year, judging from the preponderance of roots among the bundles, it will grow still farther away. In Shakespeare's England a bundle of firewood was a fardel. "Who would fardels bear," Hamlet asks, "to grunt and sweat under a weary life?" Well, the women of Tinje evidently will, or must. And how the advent of a road will change the weight and content of their fardels is anybody's guess.

From a low rise we finally glimpse Tinje, a cluster of square houses and low walls rising from a rock and gravel plain. The remote self-containment of the village gives the appearance of a windswept Shangri-La. One need only conjure its pinnacled shrines into towers and its boxy stone buildings into parapets. The flutter of prayer flags and the looming mountains require no embellishment.

The Shangri-La of legend was a Himalayan Eden, untouched by the rest of the world and free from strife and disease. Accordingly, the lives of its people were long and nearly immortal, and happiness their daily fare.

Tinje, alas, cannot say the same.

21

Feet

THE KEY to foot washing is to be personal, but not too personal. You are present, also detached. The encrusted toe before you is not your toe; it is hardly *like* your toe. So you go easy on the projections. Still, it is a human toe, connected to a human person, and it deserves a gentle touch, almost a caress, no matter if it belongs to a wizened man who smells of whiskey or a crone with one crooked tooth.

I am working next to Pete, a burly lawyer from Hawaii and former college running back. Pete has been washing feet since Ringmo. He is efficient but unhurried, beginning with a quick rinse of warm water, if we have it, but more often cold because we don't. Then he scrubs the foot with a soft yellow sponge that quickly turns gray. One hand holds the foot in an enveloping grip, while the other whisks along with the sponge. The woman who belongs to the foot looks away, at no one, but her foot is relaxed. She lets him have it. He puts it down on a clean white cobble and tenderly picks up the other. He rinses and scrubs. Next comes a light application of soap, one of the expedition's most precious commodities. Rigdzin is certain that a bag of soap was packed in one of the clinic duffels, but no amount of searching has turned it up. Perhaps it was sent to the hospital in Dunai with the errant stash of antibiotics. Mercifully, someone—I think it was Donna—has contributed a mini bar of hotel soap from her personal kit. We use it sparingly, for it may have to last through hundreds of feet. Pete and I wear latex gloves, which are almost as scarce and prized as soap, and we guard against tearing them on a toenail, a likely enough outcome given that few, if any, of the toenails of Tinje have been trimmed, possibly ever. Some have blade-like edges and ice-pick points. So we are careful. Especially when lubricated by a few suds, the latex gloves glide easily over the feet of our clients, producing giggles in some and stupefaction in others.

We kneel among the cobbles of a sandbar where the clinic tents are pitched, with our clients seated on stools before us. Others waiting their turn watch us closely. Because the wind blows hard, we weigh down everything with cobbles—the basin holding clean sponges, the rags for drying feet, and the cushions we kneel on, from which we rise periodically to stretch our backs.

Tinje is made of cobbles. The village lies at the confluence of rivers, one of which gathers its headwaters at the Chinese border. The land in every direction is glacier-cut, and in ages past enormous quantities of rock washed down from the heights, smoothing to cobbles in their violent descent. Over centuries the people of Tinje have gathered the bright stones and piled them in serpentine walls to enclose hundreds of small terraces, no two shaped alike and each scalloped into the next. The terraces ladder up the slope from riverside to village. Some of their walls stand above the earth, while others, half-buried, only hold back the ground from spilling downward.

The terraces are gardens for barley, potatoes, and buckwheat. Few are large enough to be called fields. Building the walls that separate them was an act of removal more than enclosure. The point was to get the cobbles to the side, so that a semblance of soil might be retained and plowed. The walls hold no animals in or out. Yesterday at dusk we watched a thousand village sheep and goats swarm up the terraces and over the walls on their way to nighttime quarters. Even the village yaks, which number in the hundreds, clambered freely through the labyrinth.

Pete and I are acquainted with the dung-seasoned soil of Tinje. It coats every foot like an earthen sock. Wetted, it becomes a pigment you might paint with. With every touch my gloves blacken. I try to hide this fact from my client, keeping my hands low and out of sight as I reach into a second basin for a rinse. I am not sure why I do this. Perhaps I fear that the filth on my gloves will embarrass the owner of the feet I am washing. Or perhaps it is only I who am embarrassed by the notion that there is anything wrong with chronic grime. Feet are feet, and dirt is dirt. And the dirt, including the pungent cultures breeding within it, must be removed.

I wash the feet of a man whose weathered face glows like polished mahogany but whose feet, relieved of their coating, are as pallid as mine. I wash the feet of an old woman who says the bulbous swelling on her ankle is not tender but who winces when I touch it. Clients come and go. I grow certain that none have had their feet handled, let alone gently, by another person since infancy. I observe and remove what I take to be sloughed skin along with the grime. The breeze is ever restless, yet the odor here is overripe and air-deprived. As the day wears on a scent of decay attaches to both Pete and me. In contrast to the mountain wind, which speaks of vast, eternal solitudes, this work imparts a stifling reminder of mortality.

The place where we wash feet is a kind of theater. Friends watch friends get their feet washed, and then the friends take their turn. Oddly, the women stick out their tongues when I take their bare feet in my hands. The author of the pocket language guide I carry (which, at least in my hands, is less than useless) writes that the tongue-sticking-out gesture is meant to say, "See? I have nothing in my mouth. I speak the truth!" But I have asked around and no one corroborates this notion. The sticking out of tongues seems instead to express embarrassment mixed with delight: "See how crazy these round-eyes are? And it tickles!"

This is our second day of foot washing, and I have lost track of the rest of the clinic. I hear scraps of reports: a pebble has been removed from the ear of a boy; 50 ccs of fluid were drained from a thyroid abscess; a thumb got splinted; a walking cane was "prescribed" for a man with cerebral palsy. We are running low on ibuprofen because everyone aches. We are in a world of pain, and at least half the patients receive acupuncture before seeing other clinicians. Which means they come first to Pete and me.

In Asia feet are considered impure. They live on the ground, in the dust of the field, in the shit of the sheep pen, in spring mud, in city filth. Perhaps you have been sitting a long time on someone's floor and want to stretch your legs, or you have an aching knee and want to put your foot up, but you must hold back. It is considered exceedingly impolite to expose the soles of your shoes or the bottoms of your bare feet toward another person. Yet here in farthest Dolpo we touch the feet of people who rarely or never wash them. As they live in one of Earth's coldest human habitats, they do not squander the warmth of their bodies by rinsing in ice-melt streams. Still less do they burn precious fuel to heat water for inessential bathing. They work ankle-deep in dust and mud, and because nowadays the traditional, handmade Tibetan boot is saved for ceremonial occasions, their footwear consists of low-cut Chinese sneakers. How crazy we Westerners are to care about feet!—we rich foreigners, who are rich by definition, because we have traveled halfway around the globe. How crazy this convergence between the feet of Tinje and the kneeling, white washermen!

A sustained gust sends the tents straining against their stays, walls shuddering. Loose things clatter downwind. A commotion erupts some distance behind me. People run and shout. Something important must have blown into the river. Pete and I stay at our tasks. As the hubbub continues, I feel the foot in my hands change. I cannot explain it, but I sense that the person who belongs to the foot is staring at me. I look up. She is a woman of middle age, handsome and collected in her gaze. Her eyes peer into mine warmly and thoughtfully, with no trace of shyness. It is as though I am a puzzle she might decipher. Her regard holds me. Either of us could disarm the moment with a grin and a laugh, but she doesn't, and so neither do I. I hold her soapy foot in both my hands and admire her broad, flat cheekbones, her straight black hair, and the nose ring in her septum. For long seconds she allows me into the centers of her dark eyes, where I encounter a steady, unblinking calm. I am not used to such undefended intimacy, but in this moment it feels right. We wonder at each other. Her foot is relaxed and warm. She does not flinch. The moment lengthens into two long beats, into three. The thought seems to come from her foot into my hands that in a different world we might have known each other, might have shared a laugh, a grief, an understanding. A closeness. Then a woman returning from the excitement by the river calls to her in the tone of a joke. The woman whose foot I am holding quips a reply and, looking again at me, sticks out a pink tongue. Hah! How crazy! Her grin is broad, her eyes alight. And she lets me see the light.

That night in my tent, I replay the moment in my mind. I see the woman, calm and strong. I feel her inquiry. And her warmth. I marvel at the gulf of difference between us and at the fleeting instant that seemed to bridge it. I see the woman even now, as I write this.

Weeks later we learn that word of the foot washing reached Dolpo Tulku Rinpoche, the highest lama of the region, who, although absent, is a guiding spirit of our expedition. The title *tulku* indicates that he, like the Dalai Lama, is considered the reincarnation of previous lamas of his line. People throughout the district revere him. Eventually Dolpo Rinpoche confided to Roshi that our strange practice of foot washing had an impact on the people who came to our clinics. It convinced them that we were unlike other Westerners. He said it showed that we were trustworthy. Like his compatriots, he considered the foot washing an exceedingly curious thing for us to have done, but also like them, he approved.

22

Tinje Я Us

TSEDAK DORJE has a high forehead and long, lank hair. I met him on the rutted central lane of Tinje. I did not know when we met that he was bound for our cluster of tents. He was going to dance for us that evening in a presentation of cultural arts, a gesture of gratitude by the people of Tinje for our two-day clinic by the river.

As we passed, Dorje looked at me with a direct kind of curiosity, not the usual hesitance of villagers toward foreigners. So I helloed.

And he helloed back, and stopped.

A university student home from Kathmandu, he spoke fluent English and indulged my questions about wildlife.

"Do people here often see wolves?" I asked, and Dorje said he had seen a wolf five days earlier.

"Has anyone lost livestock to wolves?" Five days ago, a man's horse was killed, he said. Probably the same wolf.

"What about snow leopards, how often do people see them?"

"Almost every month someone sees a leopard hunting blue sheep. You see, the blue sheep are often close to the village. And so we see the leopards that hunt them."

Earlier that morning, upon leaving my tent, I had been astonished to see a dozen bharal grazing at the very edge of our camp. In certain lights, their grayish coats look slate blue, accounting for their English name. Such was the light this morning. The sheep were almost luminous, and they grazed calmly and resolutely, only infrequently casting a yellow eye my way. I asked Dorje what made them so tame. Was it something to do with the national park, the boundary of which lay not far to the west?

"No," he replied, shrugging as though to say the park was unimportant. "Our lamas forbid hunting. The dharma protects the sheep."

The power of the dharma had drawn George Schaller, with Peter Matthiessen, to Shey in 1973 to study the innate behavior of bharal that were uninfluenced by human hunting. Evidently, the lamas of Tinje, like those of Shey, forbade the killing too. And the people have obeyed.

It was for Schaller, incidentally, that I asked Tsedak Dorje my questions about wolves and leopards. Schaller and I then served together on a committee, and a few months earlier he had told me that he too planned a fall trip to Dolpo. This would be his first return since '73, and he hoped to meet some of the lamas and villagers he had encountered forty-three years earlier. Alex Matthiessen, Peter's son, would accompany him, along with Sonam and Tshiring Lama, two young women, sisters, from Ringmo (Tshiring was studying wildlife biology in Kathmandu with a particular interest in snow leopards), and photographer Beth Wald, then on assignment for *National Geographic*. Schaller's expedition was led by Peter Werth, an American philanthropist and humanitarian with a long history of commitment to Dolpo. It was possible our two groups might meet, but I secretly hoped not. Compared to George's thrifty band, our village-on-the-hoof, with its scores of mules and horses, was an embarrassment of riches and a heavy weight upon the land. I wanted to avoid George's justifiable scorn for our numbers and luxury. Still, since our expedition would swing farther east than his, perhaps I could be helpful. I had asked what information I might collect on his behalf. The questions about predators and wildlife loss, which I had also asked in Tokyu and Ringmo, were the result.

The protection afforded by the dharma remained evident later that afternoon, as nearly two hundred people gathered on the tableland where we were camped (it was the site of an abandoned CIA airstrip, never used, which had been built decades earlier to supply rebels contesting Chinese occupation of Tibet).

Blue sheep grazed nearby slopes within easy rifle shot of our throng. Slowly they retreated up the mountainside, more to follow the last rays of sunset than to escape the milling humanity below. The festivities for which we had gathered included distribution of a hundred or more "Little Suns," which are solar-charged lights set in a yellow plastic frame that dangles from a yellow strap. They are the brainchild of the Danish-Icelandic artist Olafur Eliasson. The Little Suns proved handy for those of us who had used them in our tents until now—during the day we strapped them to the outsides of our packs to charge as we walked—but they were still handier in houses of stacked stones where narrow slits passed for windows and left interiors dim even at the height of day. Little Suns need no hard-to-get, soon-depleted batteries. Their light adds ease and pleasure to such mundane acts as mixing tsampa, changing the baby, finding a shoe, or just seeing the person you are talking to. We'd brought six hundred of them, the procurement of which was a modest example of Roshi Halifax's fundraising prowess, assisted mightily by Charlie McDonald and his partner, the writer Rebecca Solnit, who had accompanied a previous Nomads expedition. As a pulmonologist, Charlie well knew the deleterious health effects of indoor smoke from yak-dung fires and butter lamps. Little Suns changed at least part of that, and we freely dispensed our supply of the gadgets, giving a Little Sun to the head of every household in Tinje. Amid the steady clicking of camera shutters, each recipient was soon photographed holding aloft a Little Sun like a fresh-caught fish. The exchange of gifts continued: the leaders of Tinje now bestowed a ceremonial *kata* on every member of the expedition, each of us coming forward, one by one, to receive a scarf with head bowed.

And then a dranyan was produced. As in Tokyu, the instrument was carefully sculpted, a carved horsehead furnishing its headstock. A stool was set down for a long-faced man with dangling earrings who solemnly plucked its strings. Three others stood behind him and began to sing, while the crowd spread back to form a large circle in front of them. Tsedak Dorje and a second young man entered the circle, each followed by four young women in traditional garb. The dancers moved to the rhythm of the dranyan, the men kicking high, dipping their shoulders and spinning, their balance sometimes daring, while the women, with modest steps, turned left and then right in slow unison, never betraying a smile.

As evening shadows enveloped us, the air turned brittle. The bharal had nearly topped their long slope, tracking the last of the sun. The crowd pressed in, hip to hip and shoulder to shoulder, as though for warmth, a mass of adults and well-hugged children, the reds and blues of their quilted jackets dissolving in purple darkness. The dancers whirled and stepped. The dranyan twanged. An immense and starry sky soon spread above us, as we heard the chants of old Tibet, saw the passed-down gestures and dance steps, felt the echo of chords first struck when wild yak roamed the heights above this valley, in the days when the dharma was new.

Suddenly, a sound of motorbikes broke the spell. They were light, loud, and durable Chinese models, acquired in trade at the border post a dozen kilometers away.

Several of our group—Sonam, Wangmo, and Ranjit, a cardiologist from New York—departed the crowd to fetch their medical kits and minutes later sped away on the backs of the bikes, holding tight to the shoulders of their drivers. They later described a miles-long ride up stone-strewn paths, seeing only the cramped tunnel of their motorbike's headlamp and otherwise blind to the rough track they were on. They had no idea where they were headed. Finally their drivers delivered them to the house of a woman, old by local standards, in grave condition. She was sixty-two. They diagnosed her with a heart attack, emphysema, congestive heart failure, and pneumonia. Her history of lung issues was familiar. The woman had cooked on unchimneyed yak-dung fires all her life, breathing the smoke that filled her house. The medicos could do little, except make her comfortable.

An earlier house call produced a better result. Wendy, the invaluable Wangmo, and Becky, a photographer and wilderness medic, had visited a pregnant young woman suffering from back pain. Her mother was with her, tense and vigilant, and the windowless house felt thick with gloom. A portable ultrasound unit confirmed to everyone's relief that the unborn baby was all right. The young woman was late in her pregnancy and this was her first child: the baby might come any day. The medicos traced the woman's back pain to a urinary tract infection and gave her antibiotics and prenatal vitamins.

Toward the end of their visit, a wail suddenly issued from a dark corner of the house. It was the cry of a newborn. The clinicians learned that the young woman's forty-two-year-old mother had given birth only two days earlier. She had cut and stacked barley until three days before that. Out of shame or distraction she had not mentioned the sleeping baby. It was one of thirteen children she had given birth to, of whom eight still lived. There would have been nine, but two weeks earlier, a sister to the pregnant woman had died in childbirth along with her baby. These facts emerged as Wangmo queried the mother and the surviving pregnant daughter, whose extreme anxiety, even apart from a painful infection and impending first birth, now seemed more fully explained.

Days later and miles up the trail, word reached us that the woman whose heart was failing had died but that the younger patient gave birth, without complication, to a healthy girl. The new mother may owe her life, and that of her baby, to the serendipitous arrival of a few outsiders and a handful of antibiotics. It was a rare cure.

The mortal struggles of Tinje were not always so private. In 1959 when the People's Liberation Army swept through Tibet and the Dalai Lama fled Lhasa, thousands of Tibetan refugees poured into Nepal, many by trails into Dolpo. They arrived cold and hungry, weak and often wounded. The people of Tinje and other villages cared for them as best they could and sent them on their way southward across the barrier of the main Himalaya to cities and camps where supplies were adequate. But the waves of refugees kept coming, even as the seasons advanced. Eventually the critical passes, including Chhoila, closed with snow. Refugees who had lingered or just arrived found themselves marooned in a community possessing barely enough food and fuel to support itself. The days shortened. Temperatures fell. The cruelties of the winter gripped everyone, refugee and resident alike. Some froze. Some starved.

Hundreds of thousands perished in Tibet in 1959 and in the years of CIA-abetted guerilla war that followed.[4] And the refugees kept coming. The farmers of Tinje, who at first had been generous, were compelled to defend their meager stores of barley and potatoes with cudgels, fists, and knives. The situation was immoral. To be generous was suicide; one's own family would starve. But to refuse aid was murder, as it caused the starvation of others. The power of the dharma, to put it mildly, was strained. I learned of these tribulations secondhand, through Roshi. She said the elder who related them to her had wept as he told the tale.

A belt of limestone runs through the slopes above Tinje. It forms a cliff that is pocked with cavities. Wolves and snow leopards are said to den in some of the caves. Others once served as secret caches for grain, their entrances concealed from prowling, hungry refugees. Still others, it was rumored, were the last shelter to which the most desperate souls retreated, and from which not everyone emerged. We heard similar stories a day's walk farther north at the village of Shimen, where the same band of limestone, with similar caves, runs through the heights.

Tinje's nightmare offers a parable for the future and even for the present. It is writ large in the flow of refugees fleeing into Europe from Syria, Bangladesh, Africa, and elsewhere, people driven by war, privation, and climate change, or by all three at once, the woes of the world

---

4 A thorough and reliable account of this conflict has yet to be written. Given the secretiveness of the principal combatants, its full history may never be known. The story of Nepal's recent civil war also needs to be fully told. Known to many as a "Maoist rebellion," it lasted a decade, finally ending in 2006. Although worthy political reforms emerged from its settlement, the war's most lasting effect may be the impetus it gave to the migration of rural villagers to the cities, especially Kathmandu.

being usually intertwined. Many Central Americans and Mexicans seeking asylum in the US answer the same description. The diaspora that fled New Orleans after Hurricane Katrina was another early wave. The World Bank estimates that by 2050 climate change will have caused an additional 143 million migrants to leave their homes in Africa, Latin America, and Southeast Asia. Another estimate holds that, even if warming were limited to 2°C, as contemplated by the 2015 Paris Accords, rising sea levels will still uproot seventy-nine million people. Precision is, of course, impossible. If these numbers are half again too high, the reality remains staggering, and the estimates could just as well be too low. If the world's fever rises more than an additional two degrees, as seems increasingly likely, the agricultural foundations of vast regions will collapse, and savage heat waves and coastal inundations will render large swaths of the continents uninhabitable. With some certainty we can say that the eventual numbers of refugees will be of a scale as hard to comprehend as the immensities of geologic time. Drought, rising seas, tempests, and a host of other calamities will set whole populations in motion, making many international boundaries irrelevant and governance impossible. Such changes will elicit both the best from people and the worst. This was the case in Tinje, whose dilemma will become universal. And there will be little cure for the world's predicament, only an infinite need for care.

23

Strong Back, Soft Front

ONE DAY on the trail we met a mule with a broken leg. This was miles from Tinje and distant also in time, but the mule was a universal mule and the trail a universal trail.

Our path followed a ledge along the contour of a mountain. A copse of birches grew thickly at the widening of the path where the mule stood, creating a sense of enclosure. Rump to the mountain, the mule stared glassily at our passing line of walkers and baggage train. Its eyes, by any measure, were despondent. It placed no weight on its right rear leg. When the animal stirred—because one of us approached or because our slow-gathering crowd unsettled it—its right rear hoof waggled as loosely as a weight at the end of a rope. The cannon bone was broken clear through. The pain must have been spectacular.

It was a white mule, which means that, like other white horses and mules, it had been born gray, and over the years its coat had paled. This is a roundabout way of saying that it was no longer young. And now it was doomed.

Some of us gathered. We stood at a distance, staring at the forlorn animal, and conjectured how it had come to its predicament. We speculated about causes, but we did not engage with it. We held back as though its misery might be contagious. At least, that is what I sensed in myself. Seeing the waggle of the useless hoof made my stomach churn.

And then one of our doctors rounded the bend. Without hesitation, she dropped her pack and with a hand outstretched for the mule to smell, approached it. The mule flinched as she came near but did not try to move away. She whispered in its ear and stroked its neck. She stood close, her body lightly touching the mule's forequarters, contact the mule understood. Only a few steps behind, another colleague, a specialist in intensive care, arrived and stepped to the other side of the mule, also stroking its neck, speaking low, and standing close. The two medicos seemed to flood the brushy alcove with unhurried calm, and time began to slow.

Eventually they took the mule's pulse, which was weak and fast, and examined its mournful eyes. It was far gone, close to collapse, and there was no remedy for its fractured leg. The two women discussed euthanasia: how would they do it, and did they have the right to?

Tenzin fortuitously appeared, and in his efficient, mysterious way made contact with the mule's owner, probably the proprietor of the stone cabin and campsite not a mile behind us. Word soon came back that the mule had broken its leg three days earlier. It had neither eaten nor drunk since. And yes, the owner would not object if the mule were put down.

The women drew medical kits from their daypacks. One of them carried a few veterinary supplies. The mule's jugular vein had nearly collapsed from dehydration, but with patience and calm they managed to inject it with tranquilizer. When the mule became unsteady, they guided it to the ground. A few moments later the sedation was complete. The first woman was an ENT surgeon. She cut the mule's carotid artery with a scalpel. They stroked the mule and spoke to it in tones inaudible to the rest of us, as it bled out.

24

Lone Wolf

CHARLIE McDONALD, our medical director, is again walking without pain. After the failed ministrations of others, Wendy Lau, an ER doc, examined him. She flexed the ankle, felt the lump. Then, using an ordinary spoon to shield her thumb, she bore down on the bony mass with medical remorselessness. I don't know where doctors acquire their willingness to violate another human body. Wendy is a strong woman and she went all out. Charlie blanched. It was a though she collected all the future pain the lump would cause and concentrated it into one harrowing moment. When she released her pressure, the lump was gone. Charlie now walks with the rest of us from Tinje to Shimen.

A kilometer past the last houses of Tinje, we come to the canyon that is the route to the international border. A week earlier, Ngawong's brother would have driven his yaks up the canyon to trade Phoksundo lumber for Chinese goods. And it was down this canyon, in the reverse direction, that Tibetan refugees of the fifties and sixties fled to hoped-for safety, carrying their babies and their meager bundles of belongings.

In the hamlet at the canyon's mouth two men plow a walled-in patch of soil with a team of yaks. One guides and goads the yaks by pulling on ropes attached to their nose rings. The other leans heavily on the wooden, single-share plow to keep it from skipping out of the ground. Nearby, in another walled enclosure, a woman whacks a small mound of barley with a flail. Her tool is something like a big, hinged flyswatter. The hinge, offset from the shaft, boosts the head speed of the knotted cords that do the striking. It seems an odd and fragile design, but effective. Thwap. She strikes the golden grasses, beating the barley grains from their husks. Thwap. Soon they will be winnowed, the threshed barley tossed in the air, the lighter husks blown away, the grains falling to the ground to be collected. Thwap. Thwap. The woman pauses. She gives us a passing glance as we walk by. Thwap. A hill of barley, awaiting attention, looms behind her. It is twice her height.

Time has settled a heavy grace upon the land. It is doubtful that the grain harvest, or any tool or practice within it, has changed in centuries. Equally, the land feels sacred. Even after we leave the houses behind, we are never out of view of a shrine, not for kilometers. We see them in every direction. They stand by the trail, or on knolls above us, or on ledges far up the canyonside, each one a tower of plastered stone, set on a squat base. The shrines are sentinels. They are sculptures. They are messages and reminders. Roshi has explained that buried in the core of each tower is a pole on which prayers are inscribed in blood-red ink. Offerings of sacred essences surround it, and the ashes of a revered lama may lie entombed there as well. The shrines are called *chortens*, the word being the Tibetan equivalent of *stupa*, which is Sanskrit. They are built according to a precise formula, the elements of which—does this sound repetitive?—symbolize aspects of the Buddha's mind.

Pairs of concentric circles, some painted, some daubed in colored plaster, decorate the base of the chorten. These are the Buddha's eyes, which survey the world and every passing traveler. One walks by a Buddhist chorten on the left, as though to circle it in a clockwise direction. But a Bön chorten should be passed on the right. It is a mystifying distinction. A Buddhist chorten looks the same to me as a Bön chorten, and so if there is anyone who might know one from the other, I follow them, and if there is not, I generally bet on the Buddha and pass on the left. Even so, I wonder, if the chorten is Bön, to whom belong the painted eyes that watch me?

Now the trail descends a defile between contorted walls. The strata warp and curl, red formations crushing black ones, rocks upthrust, wrenched, and fractured. One feels the echo of prodigious violence. The trail threads through mounds of rounded white boulders as big as buses. The floods that might have tumbled those colossi strain imagination, but no living trace of them remains. Only a single, pattering waterfall leaks down the desolate walls.

At last the defile opens into a broad canyon. Somewhere in the canyon, Prem and Tenzin saw a wolf. It was large and white, perhaps the same one Tsedak Dorje saw close to Tinje. It loped across the trail in front of them and continued up the opposite canyon wall. They watched it climb with easy effort, never slackening pace, as though gravity did not weigh on it. High up the slope a band of blue sheep grazed. The wolf raced past the little herd, as though indifferent. But above them, it quickly turned. The wolf charged down upon the sheep, which scattered in a cloud of dust. When the air cleared, all the animals were gone. Tenzin thought probably the wolf had killed. Prem was not so sure.

News of the sighting makes me envious. How I wish I had been with Tenzin and Prem to witness the wolf, to feel it charge the landscape with energy, to watch it race up the mountain in defiance of gravity. But on reflection, my envy turns to sorrow. A single wolf sends a gloomy signal. Wolves are social and familial. Their unit is the pack. Of their current status in Dolpo, Schaller explains, "Wolves are usually seen singly or in pairs, rather than in small packs as expected during the season of our trek, suggesting that the pups had been killed in their den by villagers in retaliation for predation on their livestock, which all too often is left unguarded." A single wolf in autumn is yet another symptom of broken nature, no different here than on my home ground of New Mexico, where coyotes are routinely poisoned, trapped, or shot on sight and the limited reintroduction of wolves, after a long period of extirpation, has met continuous opposition from livestock operators. With their respective livelihoods at stake, relations between humans and their fellow predators are always fraught—delicate on the rare occasions when a balance is struck and almost invariably injurious to Sapiens' competitors when one isn't.

My thoughts—on predators, poisons, and leg traps—carom from one doleful reflection to another. Then I catch myself and laugh at finding so much fault with the present, even when the present brims with wonder.

Later in the day we pass a group of Swiss trekkers who say that they saw the Schaller, Werth, and Alex Matthiessen group a few days earlier in Saldang, the site of our next clinic. But our paths will not cross. George and the others departed southward from Saldang on a direct route to Numa La and Bogu La, not eastward toward Tinje and Shimen.

Over a long and astonishing career Schaller has studied virtually every large, charismatic, terrestrial mammal: gorillas in Rwanda, lions in the Serengeti, tigers in India, jaguars in Brazil, brown bears in Tibet and Alaska, and snow leopards throughout the Himalaya. Not long ago, we chatted about the progress of conservation during our lifetimes. "At least we've won a few battles," I ventured.

"Don't kid yourself," he replied. "Those weren't wins. They were just temporary stalemates." Schaller knows as well as anyone the state of things, yet he persists.

25

An Entire Heaven and an Entire Earth

FOR EVERY five wild animals that existed in 1970, there are now two. So says a 2016 report from WWF International and a host of partners, drawing on long-term monitoring of 14,152 populations representing 3,706 vertebrate species. The numbers are big, but the data are imperfect, being drawn unevenly from around the globe. Moreover, the species selected are not a systematic sampling of Earth's biodiversity; they simply reflect what scientists happened to be studying forty-some years ago and have managed to continue to study. But the data make a point, and, for me, the point feels personal. I was in college in 1970. The colossal, 60-percent decline of the wild world has occurred during my adult life, and it continues unabated.

Every generation in the modern era has witnessed the depletion of wildness. In March 1856, Henry David Thoreau mused in his journal about the wildlife of New England: "When I consider that the nobler animals have been exterminated here . . . I cannot but feel as if I lived in a tamed and, as it were, emasculated country . . . I should not like to think that some demigod had come before me and picked out some of the best of the stars. I wish to know an entire heaven and an entire earth."

In 1922 Aldo Leopold explored what seemed to him an entire heaven and earth in the verdant, little known delta of the Colorado River in Sonora, Mexico, just below the US border. The wildness was not as devoid of human imprint as he thought, for upstream diversions from the river had already forced the Cocopah natives to move away. Even so, the "milk and honey wilderness" that enchanted him did not last. Twenty-two years later, he wrote, "I am told the green lagoons now raise cantaloupes. If so, they should not lack flavor." His reflection closes with deep nostalgia: "Man always kills the thing he loves, and so we the pioneers have killed our wilderness. Some say we had to. Be that as it may, I am glad I shall never be young without wild country to be young in. Of what avail are forty freedoms without a blank spot on the map?"

The ruefulness of Thoreau and Leopold would be different today. Neither of them could have anticipated the pace and scale of planetary diminishment that is now commonplace. A 2017 paper in the *Proceedings of the National Academy of Sciences* found that "Earth is experiencing a huge episode of population declines and extirpations" among terrestrial vertebrate animals. The losses, say the authors, are tantamount to "biological annihilation." These are not the usual bloodless scientific terms. With some passion, the authors assert that the current rate of extinction, about two vertebrate species per year, is a hundred times greater than the background rate of the past two million years.

Even so, most accounts of the ongoing assault on wildlife hide the breadth of current losses, which extends, not just to species that are rare and endangered, but to shrinking populations of creatures that are widespread and, at least for the present, common. The investigators looked at roughly half of all mammals, birds, reptiles, and amphibians (27,600 species). They found that a third (8,851 species) are experiencing "declines and local population losses of a considerable magnitude." They further assert that extinctions of local populations are "orders of magnitude more frequent than species extinctions." If these losses presage the extirpation of species, they conclude, the cumulative trend is producing "planetary defaunation."

This assault on nature is hard to face. If the winnowing of biodiversity by humans can be seen as an evolutionary force, it constitutes a most unnatural selection. Yet we of the present time are called to bear witness to it. Akira Kurosawa, the visionary director of such classic films as *Rashomon* and *Seven Samurai*, said he owed much to a lesson taught him by his older brother Heigo. When the Great Kanto earthquake of 1923 leveled much of Tokyo, Heigo took Akira into the city and forced him to look at the carnage. Human corpses and animal carcasses sprawled everywhere, in every posture of death. Heigo, then seventeen, forbade thirteen-year-old Akira to look away. "If you shut your eyes to a frightening sight," he explained, "you end up being frightened. If you look at everything straight on, there is nothing to be afraid of."

The brave moral of this story, unfortunately, is undercut by Heigo Kurosawa's suicide ten years later.

26

Shimen

IN SHIMEN our camp fills four paddock-size barley fields, so small are the fields, so many are we. Needing solitude, I go to an empty paddock a distance away. Only wisps of straw remain from harvest. The gray, fine-grained soil is like a powder of cement. The farmer of this patch has hoed up a series of parallel berms across the field, inches high, so that the plot looks to have a ribcage. The berms spread the irrigation water, each tier metering water to the next. For forty years and more, I have irrigated my own fields and witnessed irrigation by others, but I have never seen such careful precision. The treatment of the ground in Shimen seems closer to sculpture than to farming.

Across the valley Govinda hazes a dozen mules up a pencil-thin trail. Govinda stands out from the other muleteers as he wears his straight hair very long—it reaches to the middle of his back—and he moves with a feminine, balletic grace. With every glance and gesture he also radiates a fierceness that gives pause even to Prem. And now he is angry. Word has passed that the headman of Shimen has charged a thousand rupees (about nine dollars) per animal for overnight grazing, an extortionate rate that Govinda and the muleteers now following him have refused to pay. Instead, they are climbing a thousand feet and will trek a mile or more to graze their animals outside the precincts of the village. Such an errand, doubled by the return, quadrupled by the fetching in the morning, would be a day's work for most of us.

The steep trails and meticulous fields, to say nothing of the crumbling hermit retreats perched on ledges high above the village—every feature of this place testifies to hard and diligent labor. Entering the village we passed a massive wall of carved stones, said by Prem to be the second-longest mani wall in Nepal. A *mani* stone is a stone engraved with a sacred image or mantra, usually "om mani padme hum." (The stone borrows its name from the "mani" in the mantra, which is sometimes translated as "the jewel is in the lotus." That rendering, however, is only a beginning; deeper interpretations of the mantra embrace a universe of meaning.)

A smooth flattish cobble will bear a decipherable message for centuries. People carve mani stones for departed family members, believing the merit thus earned will ease their loved ones' transition to the next life. Sometimes stones are carved for an especially valued horse or yak. The departed souls of tiny Shimen, over countless years, must include multitudes of humans and beasts, for the mani wall we passed was hundreds of meters long, wider than a country lane, and nearly as tall as a person. It was religion become geology, a rock ridge to weather the ages.

Like Tinje, Shimen presides over a mosaic of hundreds of stone-walled fields. From a ridge above the village I count thirty terrace levels rising step by step from river gorge to mountain wall. A stream, now dry, exits a cleft in the wall. In spring, filled with snowmelt, it must be a torrent. Irrigation ditches split off from the stream channel, rounding the tops of the fields and splitting repeatedly until their capillaries deliver water to every farmed patch. The water delivered by the ditches is vital early in the growing season and, as snowmelt tapers off, farmers depend on remnants of the monsoon to straggle past the Himalayan wall and nourish their crops. And after the monsoon comes harvest. At least, that is how things are supposed to work. The timing is delicate. In a warming world with earlier snowmelt and a more variable monsoon, farming goes from a gamble that is regularly won to an increasingly bad bet.

Apportioning water across a matrix of innumerable fields demands wisdom and cunning. Historians say the first mathematics arose in Mesopotamia and Egypt to accomplish the task. One must measure the water and time its use so that it flows through and between families and clans, be they friends or enemies. Although the supply of water may be limited, the potential for cheating and argument is not. One must also clear rubble from the ditches, repair inevitable breaches, and rebuild the diversions when the torrent washes them away. Assembling the necessary labor, deciding who directs it and how its burdens should be shared, becomes basic to the politics of survival. The "simplicity" of village life is anything but.

27

Yatra

THE LONG climb up from Shimen, which Govinda took at a lope, is for me a metronomic trudge. Step upon step, breathing hard, I think only of the act of walking. With amusement, I realize that I am still learning to walk. I am learning, for instance, that I must walk slowly enough that I do not need to rest. At the right pace, my lungs and legs work in synchrony, and I neither gasp nor feel weak. If I stop longer than a few seconds, I lose my rhythm. It is as though my lungs and legs leave their posts, go on break, and refuse to come back. And even when forced to resume their duties, they resist working together. The trick is to keep them at their jobs. Like the tortoise that raced the hare in Aesop's fable, I need to stay in motion. Old Aesop must have walked up many a mountain.

I am lightheaded near the top of the climb. My trailmates sprawl on rocks, their packs beside them. And here comes a family of villagers, mother, father, and young boy, headed down to Shimen. In Dolpo, upon encountering strangers, you bow slightly. If your hands are free, you might bring them together in a gesture of prayer. And you say, "Namaste." It is a simple word but profound in its depth. It roughly translates as "I bow to the divine in you." By contrast, our English "hi" and "hello" have no more meaning than marks of punctuation.

Each member of the passing family says "namaste" to me.

"Namaste," I say back, and their bright smiles continue to shimmer in my mind's eye minutes after they're gone. I probably need to drink some water.

This morning Tenzin spoke in our circle. As usual, his delivery was staccato and epigrammatic, more poem than paragraph. Tenzin's gift is to mix the cosmic with the trivial. "Stop. Take view into you," he coached. "Absorb mountain. Eat later. More important than snack bar, than granola bar, is getting blessings from our journey." He paused, looking around the group to gauge his audience. "Every day a yatra," he said, a *yatra* being a religious journey or pilgrimage. "All we have been through, you will feel this. You will feel lighter, I think."

Such lightness, I imagine, would render one indifferent to whether the trail climbed or fell, whether the distance to go was long or short. Such lightness might even enable the resolute pilgrim to round a bend and say "namaste," and mean it, to any other creature, be it saint, songbird, or snow leopard.

On this day, alas, I cannot tell my lightness from my lightheadedness. But, perhaps, if I keep walking, eventually I might.

28

The Grandeur Sentence

THE WORD "evolution" does not appear in the first edition of Charles Darwin's *On the Origin of Species*. Nor does its cognate "evolutionary" or any similar word, except one. The verb "evolve" occurs once. It appears in past tense as "evolved," and its placement is significant. It is the last word of the last sentence of the last chapter of the book.

Darwin gestated his final sentence a long time. A version of it concludes his first attempt to commit his theory to writing. This was a thirty-five-page "sketch" he drafted in pencil in 1842, almost six years after he returned home from the *Beagle* and still seventeen years before the 1859 publication of *On the Origin of Species*. He was thirty-three years old. It is an ambitious sentence in which he attempts both a summation and a defense of the new worldview to which his reasoning has led him. His phrasing is labored and verbose. His thoughts crowd each other, but he writes with conviction and urgency. We will get to the final, elegant version of the sentence in a moment, but, for now, here is the first draft as it appeared in the sketch of 1842:

> There is a simple grandeur in the view of life with its powers of growth, assimilation and reproduction, being originally breathed into matter under one or a few forms, and that whilst this our planet has gone circling on according to fixed laws, and land and water, in a cycle of change, have gone on replacing each other, that from so simple an origin, through the process of gradual selection of infinitesimal changes, endless forms most beautiful and most wonderful have been evolved.

Clearly, the sentence needs work. This early version, however, makes plain Darwin's concern for the criticism that will be directed at his theory and at him once he makes his ideas public. The objections will take many forms, but the central thrust will be that Darwin has debased human life and drained it of the grandeur that the Creator intended for it.

The established view in Darwin's day was that species were immutable: animals were animals, each one everlastingly separate from every other, and man, whom God had created in his own image, was his masterpiece. No organic connection between mankind and the animal kingdom existed. (Unfortunately, women in that era were commonly viewed as an imperfect, if necessary, adjunct to man.) In the worldview of English-speaking people and of Western civilization generally, the Creator had endowed man with godlike reason and a soul, neither of which animals possessed. Many non-Western traditions (but by no means all) shared this view, asserting that the gulf between humans and animals was unbridgeable and that to deny this separation, as Darwin would do, constituted a philosophical error and a religious sin. Early on, Darwin began to prepare his defense. He would argue that although the organic connection existed, although *humans were animals*, grandeur nevertheless remained.

He said it was grand that the planet circled on according to fixed laws that Copernicus, Newton, and others had divined. It was grand that land and water, in ways that Hutton, Lyell, and other geologists had described, cycled through endless changes. And it was grand that the biological creatures of Earth, by virtue of incremental changes that were at once minuscule and mysterious, had developed into "endless forms most beautiful and most wonderful." Darwin entreated the world to understand that his theory did not debase life. Life remained *grand*, and it remained a fit testament to the wisdom and generosity of the Creator, even when, perhaps especially when, its most subtle workings were understood. Rather than lessen humankind's admiration for the miracle of creation, such an understanding should, as he wrote in the penultimate sentence of his sketch, "exalt our notion of the power of the omniscient Creator."

Darwin knew that his theory would shake the world, not least because it shook him. And he knew his critics would not soon or easily accept his arguments about grandeur and exaltation because he himself still struggled to believe them. He was a divided man. He lived on inherited wealth. His enviable social status as a gentleman of leisure, which enabled his unfettered pursuit of science, depended on the established order. Only radicals and revolutionaries (numerous in industrializing nineteenth-century Britain) denied that a divine hand lay behind the organization of life and thereby justified the present social and political structure. To say that species arose from entirely natural forces pulled a brick from the foundation on which everything of value to him was built: his wealth, his status, the respect of his peers, and even his domestic tranquility, for his wife, Emma, and most of his relations were firmly traditional in their views.

Through the most productive years of his life, Darwin kept his ideas secret. Surely the concealment ate at him. Some say it contributed to the gastric misery that dogged him most of his life. His illness began shortly after his return from the *Beagle* and continued for four and a half decades until, having weakened his heart, it laid him in his grave. In all those years he never experienced a sustained period of normal good health. Instead, he suffered fits of vomiting. Continuous nausea. Paralyzing headaches. Outbreaks of facial eczema that left him unrecognizable. Insomnia. Boils. Spasms. A swimming head. Depression. Spots before his eyes. Raging flatulence. Trembling. Twitching. The threat of fainting. Stomach pains. Cold shudders. Heart palpitations. And more vomiting, always the vomiting.

On occasion his ailments prevented him from working for weeks at a time. Typically they limited him to a few hours a day. When the American botanist Asa Gray and his wife, Jane, visited the Darwins at their home in Down, England, in 1868, Jane noted that Darwin, then only fifty-seven, was "entirely fascinating," but "his face shows the marks of suffering and disease . . . He never stayed long with us at a time, but as soon as he had talked much, said he must go & rest, especially if he had had a good laugh."

Evidence suggests Darwin may have contracted Chagas' disease from insect bites in Patagonia. And effects of the Chagas parasite may have been further exacerbated by an infection of *H. pylori* bacteria (which our clinic was finding to be widespread in Dolpo). These plagues could have accounted for the majority of his symptoms, but, to be sure, he also suffered from inner conflict. The stress of the approaching publication of *On the Origin of Species* in 1859 coincided with a series of especially debilitating crises.

Darwin's challenge, one of many he faced, was to convince his readers that the grandeur of human existence remained inviolable, even as that existence became subject to the coarse facts of biological survival and the contingency of evolution. Prior to Darwin, humankind's tenancy of Earth was considered to be part of a divine plan, a destiny that was purposeful. Life was headed somewhere. To the Second Coming. To the City on the Hill. To a Utopia of Progress. Choose as you will. There was a grand scheme, and its elements were intentional. Now Darwin was saying that the future had no preselected destination. Powered by wholly material forces, evolution would go on and on forever, directed toward no assured end and obedient to no assured purpose. "I know how much I open myself, to reproach, for such a conclusion," Darwin wrote, "but I have at least honestly & deliberately come to it." And honestly and deliberately he presented his fully evolved view in the final sentence of *On the Origin of Species*. The sentence was much improved over its 1842 version and deserves its place as one of the most brilliant statements in the literature of science. Many would go further to argue that no qualifier is necessary; it is simply one of the greatest sentences in the literature of the world:

> There is grandeur in this view of life, with its several powers, having been originally breathed into a few forms or into one; and that, whilst this planet has gone cycling on according to the fixed law of gravity, from so simple a beginning endless forms most beautiful and most wonderful have been, and are being, evolved.

Darwin's conciseness expresses his hard-won confidence: "A simple grandeur" has become simpler; it is "grandeur" alone. Perhaps more interesting than any deletion is an insertion: before the final word "evolved," Darwin adds, "and are being." He is saying that Creation is not finished, that endless forms are still forming, that the first week of Genesis is still the week in which we live. This apostasy, notwithstanding its truth, was one of many points that laid him open to accusations of atheism, and in the second edition of *On the Origin of Species* he mildly defended himself by inserting after "originally breathed" the words "by the Creator."

He continued to tinker with *On the Origin of Species* through six successive editions, correcting errors, refuting criticisms, adding new evidence. Words that his ideas had made commonplace—"evolution" and even "evolutionist," plus many more uses of "evolve"—find their way into the text. In the final edition (1872) he drops the tentative "On" from the title and leaves the book to be known simply as *The Origin of Species*.

The debates launched by his masterpiece continue, and certainly the question of grandeur remains problematic. Few among us are immune from doubt about the importance of humanity's place in the cosmos. Here on the busy playing field of Earth, the glorious achievements of human beings are plain to see, never mind the ugliness. Shakespeare is there. And also Bach and the Buddha, Socrates, Tolstoy, Lincoln, Mandela, Mother Teresa, Marie Curie, George Eliot, Gandhi, and, yes, Darwin. Uncle Walt is there too, along with the rest of the aunts and uncles, but of course this kind of list-making soon becomes absurd, as it mainly illuminates the limitations of the list-maker's imagination and education, mine included. No one can adequately enumerate the noteworthy among the human tribe, and still less the legions of decent and admirable people who answer George Eliot's description of her heroine, Dorothea Brooke, in the final words of *Middlemarch*: "The effect of her being on those around her was incalculably diffusive: for the growing good of the world is partly dependent on unhistoric acts; and that things are not so ill with you and me as they might have been, is half owing to the number who lived faithfully a hidden life, and rest in unvisited tombs."

But if we soar to an altitude of thirty thousand feet, or higher still, and look at Earth from the emptiness of space, as Darwin implicitly taught us to do, the view changes. From a distant vantage, our perspective becomes dispassionate and ecological. Our planet—a cloud-wreathed blue gem hanging in the great void—remains incomparably beautiful, but human particularities slip away. Time, travail, and great ideas dissolve, leaving only Petri Earth, where Sapiens swarm toward the limits, copulating, birthing, dying, struggling to survive, and in the aggregate seizing and exploiting nearly every scrap of space and energy they can command. Human armies stream across the land like masses of ants, forward and back, leaving devastation in their wake. It is hard to see grandeur in the horrific suffering of Aleppo, Fallujah, or Tinje, inflicted by Sapiens on Sapiens. It is equally hard to see grandeur in Petri Earth's plundering of the natural world or in the consequent upwelling of loss among all living things.

29

The Grandeur of Saldang

WE'VE BEEN twenty days on the trail, or twenty-one. I've lost count, not that it matters. We have reached Saldang, the largest settlement in Upper Dolpo. As we approached, the country seemed to rise and open up. Mountains no longer hemmed us in; we had attained the heights. Now we can see range after range, stretching outward for twenty, forty, sixty miles and more. Our view extends into the Tibetan Plateau. The sierras slice the distance in waves, bald peaks like whitecaps on an infinite sea. Dwarfing all is the sky above, a dome of cobalt slowly fading to the blue-black of night.

The final climb to our campsite was a wicked joke. From river to ridge, Saldang spreads up twelve hundred feet of precipitous mountain slope, and our campsite lay at the very top. Down by the river, before our climb, we were already tired and hungry, yet now we faced the greatest effort of the day. When at last we reached at our aerie at the bottom of the sky, we dropped to the ground and drank the last of the water we carried, hardly moving until a chill wind began to stir. Our tents spread across a rare patch of level ground. Below us the houses of Saldang cling to the giddy slope like swallows' nests. Built of the same rock and soil as the mountain, they are distinguished from the weather-smooth roundness of the land only by the geometry of their right-angle walls.

We see our human places as analogs to the natural world. The houses of Saldang look like clusters of birds' nests. The village is a beehive of activity, an anthill of organization. In Tierra del Fuego Darwin was shocked by what he perceived as the apparent closeness of the natives to a purely animal existence. This was not a brief or lightly held impression. Between 1832 and 1834 the *Beagle* paid three substantial visits to Tierra del Fuego, and each time Darwin observed the native Fuegians with care.

We need to pause here. There is no question that Darwin described the natives of Tierra del Fuego in terms we understand as racist today. His belief in his own superiority and that of his nation and culture was clear. So was his condescension toward the Fuegians. His attitude, while typical for his time, is unacceptable in ours. That much is sure. But we must also acknowledge that precisely in this locus of conflicting values Darwin's genius kicks in: although he saw the Fuegians as living almost like animals, this perception did not provoke in him a condemnation. Instead, it caused him to look at things with the terms reversed. For Darwin, the nearness of some people to the animal world suggested that all people had once belonged to that condition. In the context of his time, Darwin had begun to think the unthinkable—that even Englishmen were animals, that all humans, at a fundamental biological level, were animals. This was apostasy in his nineteenth-century world, and he would suffer much for embracing it. But he held to his view, and today we are the better for it. We owe Darwin thanks for teaching us—all of us—to see ourselves as part of the fauna of the world. Because of him we appreciate more fully the herd behavior of a crowd, the preening before the prom, and the alpha-male sparring in the executive suite and the locker room. Who can doubt, at an opera gala or a block party, that the resplendent outfits worn by beauties of all sexes are not analogous to breeding plumage?

But perhaps such a view still looks at things the wrong way. The coyote flips a twig in the air and catches it joyfully. Is this not play? The bison and the elephant prod the bodies of their dead and stand sentinel over them. Are they not mourning? Having given birth, the humpback whale attends her baby as solicitously as any human mother.

The infant and the puppy roll together on the rug, the glee of one not different from that of the other. However much we are like them, they are like us, all cousins sharing descent, not from Adam and Eve, but from clusters of cells that commenced the generation of all life, from maple trees to marmosets and from bacteria to Beethoven. We share 92 percent of our genes with other mammals and nearly a quarter of them with yeasts. Within the vastness of the universe, life on Earth is a small and close society. This was the grandeur that Darwin saw, and in it he found consolation. It was the grandeur of the web of life.

30

Galapagos

IN THE last days of 1831, when Charles Darwin boarded the HMS *Beagle* to begin the voyage that would take him around the world, the hesitant twenty-two-year-old believed in the immutability of species. He believed that animals existed in their present forms because, through divine or other means, they had been created that way. He believed that their forms had not changed since their first appearance on Earth and that they never would.

Roughly five years later, not long after the *Beagle* hove into Falmouth harbor on October 2, 1836, and he again trod English soil, Darwin began outlining a profoundly different view. He'd come to believe that species were somehow plastic, susceptible of alteration. He'd absorbed Charles Lyell's uniformitarianism—the idea that forces still observable in the present had gradually and continuously altered the surface of Earth—and he suspected that a similar process of incremental, additive change might also have shaped plants and animals. Given sufficient time, might it not be possible for a line of giant armadillo-like creatures to become shrunken versions of themselves, small enough to be carried in a grain sack? Or might lizards with round tails living on dry land share a common ancestor with flat-tailed lizards that swim and forage in the sea? He did not yet have a theory to explain these relationships, but, as he later confessed, "the subject haunted me." He called the hypothesized shifting of animal forms "the transmutation of species."

Darwin's family, friends, and biographers have exhaustively debated his change of view. Was it slow or sudden? Did he draw his ultimate conclusion in a gradual fashion as the voyage of the *Beagle* progressed, or did it come to him back in England, when he reflected on his experiences and studied his collections? Darwin's own recollections are ambiguous and sometimes contradict his earliest notes, but in his *Autobiography* he mentions being "deeply impressed" by three lines of evidence. The first concerned the fossils of South American animals that were armored like modern armadillos but many times larger. The second was the phenomenon of similar species (different varieties of rheas in particular) succeeding one another as he progressed down the continent, each species dwelling in a somewhat different climate and habitat. And the third included the multitude of revelations afforded by an equatorial island chain, previously known only to buccaneers, whalers, and a few infrequent explorers, where the *Beagle* lingered for five weeks in September and October 1835.

The Galapagos archipelago consists of eleven major islands and numerous islets and smaller features, all of volcanic origin, scattered across twenty thousand square miles of sea. Darwin quickly realized that the islands were young, formed by geologically recent volcanism, previous to which "the unbroken ocean was here spread out." As a result, the life-forms of the islands—the plants, reptiles, nonpelagic birds, and a single native mammal, a mouse—were young too, and on this archipelago "we seem to be brought somewhat near to that great fact—that mystery of mysteries—the first appearance of new beings on this earth."

Darwin's Galapagos experience stretched through days of exploring, gathering, making notes, and shooting. Some of the shooting may have been for sport, but its main purpose was to enlarge his collections of taxidermic birds and other creatures. Captain FitzRoy and other shipmates also hunted specimens for Darwin, which he inspected and hastily labeled. He struggled to keep track of where the samples came from and felt harried by the need to assure their preservation and proper storage. Inevitably, time ran short. "It is the fate of most voyagers," he wrote, "no sooner to discover what is most interesting in any locality, than they are hurried from it."

In *The Voyage of the Beagle*, Darwin writes about the Galapagos differently from the way he records his experiences in other places. He provides little information on day-to-day activities and instead discusses the islands' flora, fauna, and other features in summary. One might contend that this was a strategy of economy by which he compressed five supremely busy weeks into fewer pages than a day-to-day diary would have required. Fair enough. But equally persuasive is the idea that the Galapagos *made sense* only in summary, that the connections between observed facts were so numerous that a synthesis of them, rather than the individual facts, became the story.

One connection was the obvious relatedness of the biota of the islands to the flora and fauna of the South American continent, six hundred miles away: "It was most striking to be surrounded by new birds, new reptiles, new shells, new insects, new plants, and yet by innumerable trifling details of structure, and even by the tones of voice and plumage of the birds, to have the temperate plains of Patagonia, or the hot dry deserts of Northern Chile, vividly brought before my eyes."

Another realization involved island-by-island differences among the animals and plants of the archipelago: "I never dreamed that islands, about fifty or sixty miles apart, and most of them in sight of each other, formed of precisely the same rocks, placed under a quite similar climate, rising to a nearly equal height, would have been differently tenanted." Assuming that ancestors of those "tenants" arrived on the Galapagos in the recent geologic past, they must have then spread to the various islands, where over the course of time they developed differences: "Several of the islands possess their own species of the tortoise, mocking-thrush, finches, and numerous plants, these species having the same general habits, occupying analogous situations, and obviously filling the same place in the natural economy of this archipelago." So why should differences have arisen, island by island? Darwin did not know—yet. But, he wrote, "This circumstance strikes me with wonder."

The finches of the Galapagos may be the most famous birds in the history of science. Ironically they caused Darwin much embarrassment. He collected six types of finches from the islands, and he thought their differences intriguing, but he did not initially view them as significant. As a result, he failed to label them according to the islands from which they came—an omission he would greatly regret. He did better with mockingbirds, or mocking-thrushes, as he called them. In their case he knew he was dealing with multiple species from the outset, and he cataloged them carefully, soon recognizing that all the specimens from Charles Island represented a single species and that those from Albemarle Island did likewise. The intrigue heightened when he realized that the specimens from James and Chatham Islands together constituted a single species, the crucial factor being that James and Chatham Islands were connected by smaller islands that lay like stepping stones between them.

As to the finches, when he returned to London, he entrusted his collection of bird skins to the Zoological Society of London, where he could count on the multitalented John Gould to sort the finches from the grosbeaks, the wrens from the blackbirds, and every other kind from the rest. Gould was a busy man, much in demand, and he was not able to attend to Darwin's collection immediately. When he did, the results were stupefying. Every grosbeak, wren, and blackbird in Darwin's collection turned out to be a finch. Altogether, his finches totaled thirteen species, not the six he thought he had identified. The birds came in different sizes and their beaks reflected intricate adaptations for cracking open the seeds of a wide range of plants. Their resemblance to grosbeaks, wrens, and blackbirds underscored the diversification they had undergone, which enabled them to exploit habitats that on the continents were occupied by birds of different genera. Evidently the Galapagos finches, which all belonged to a single genus, shared common ancestors—"Adam" and "Eve" finches that had arrived on the islands in a past age—yet their "transmutation" into separate species must have been geologically recent, for it could not have antedated the formation of the islands.

Darwin found the behavior of Galapagos birds as curious as their diversity. They had no fear of humans. "A gun here is almost superfluous; for with the muzzle I pushed a hawk off the branch of a tree." He was able to kill small birds, and even a buzzard, by swatting them with a switch, and sometimes "with a cap or hat." He likened their tameness toward humans to the manner with which "in England shy birds, such as magpies, disregard the cows and horses grazing in our fields," and he observed that the Galapagos birds, even when "persecuted, do not readily become wild." Darwin eventually encountered reports of similar tameness among the birds of other isolated islands including the Falklands and Tristan da Cunha, where human presence was still early and light. His assessment of this trait foreshadowed his future thinking about natural selection: "We may infer from these facts, what havoc the introduction of any new beast of prey must cause in a country, before the instincts of the indigenous inhabitants have become adapted to the stranger's craft or power."

On the Galapagos island of Española there exists a rookery of waved albatrosses (*Phoebastria irrorata*), the only albatross of the tropics. Not having wings adapted for rapid flapping, these masters of the sky, possessing a wingspan that can reach eight feet, need help getting airborne, which they obtain by essentially jumping off a cliff into a strong wind. As Española provides the necessary cliff, albatrosses lay their eggs and raise their young there. Darwin's experiences in the Galapagos and the collections he made from the islands had led the young scientist to an analogous precipice. He had run out of room for simple walking. Would he finally leap and fly?

31

Invisible

ON A trip to the Galapagos in 2012, I visited Española's cliffs to watch the albatrosses launch. I lagged behind my group, which was out of sight, and ambled up a guttered path through desert scrub. Suddenly the underbrush beside me shivered, and an albatross waddled out. It was a big bird, taller than knee-high, with a long, thick, hooked yellow beak. I stopped. The bird walked up to me and stopped but seemed not to see me. Had either of us taken one step more, we would have collided. I watched for signs of anger or arousal, lest the bird slash my bare shins with its formidable beak. But it was entirely calm—oblivious, really. The bird halted in the center of the trail, one waddle from my foot. I might have patted its creamy head. It showed no more interest in me than in the cactus that fringed the trail. The albatross ceremoniously fluffed its wings and settled into the ground. There it sat, not moving. I remained still too. A minute passed, then two, perhaps three. The albatross made no sound. Suddenly it rose slightly and ruffled its feathers anew. It waddled a half step rearward and looked down. Gleaming between its webbed blue-gray feet was a large, bright egg. The albatross tapped the egg twice with its beak, then with seeming satisfaction, shuffled forward, reared slightly back, and sat on the egg as though on a throne.

The bird seemed well pleased by events, and I had no desire to diminish its happiness. Even so, I am not sure I had the power to disturb it, short of striking it, for the albatross paid me no notice. I retreated a step and detoured through the scrub, leaving the albatross to brood its egg.

Every creature, human and albatross alike, filters the information of its senses. Or rather, our senses are tuned to the information we most need, so that superfluities go unnoticed. George Eliot wrote, "If we had a keen vision and feeling of all ordinary human life, it would be like hearing the grass grow and the squirrel's heart beat, and we should die of that roar which lies on the other side of silence. As it is, the quickest of us walk about well wadded with stupidity."

Such stupidity spares us being overwhelmed, our senses fried and our nerves wrung out by too much stimulation, but how enlightening, although possibly frightening, if for a moment our wadding fell away and we glimpsed the world entire. Perhaps that is what psychedelics or the deepest meditation enables us to do, but perhaps, even then, we only scratch the surface of the actual.

32

Mechanism

SCIENCE NEEDS to say not only what happened but how it happened and, if possible, why. Wegener's theory of continental drift was essentially descriptive; it presented a scenario for a massive probability, but the evidence it mustered was largely circumstantial, and in any case, not much of the scientific world was disposed to listen. Critics asserted that Wegener could not demonstrate the actual fact of the continents' movement or fully explain what caused it. He only presented information indicating that Earth's landmasses *must have* changed position over time.

Darwin faced a similar predicament. His evidence for the "transmutation of species" included the mute testimony of the finches, tortoises, iguanas, and mockingbirds of the Galapagos. He could cite fossils that were linked to modern species but also clearly different. He could summon volumes of human experience in selective breeding, producing many domestic strains of sheep, cattle, dogs, goats, pigeons, and other animals. But he could not name the force in nature that was analogous to the human horse breeder who develops a line of thoroughbreds or Clydesdales. The horse breeder is an agent of change. He selects the animals that will produce the progeny bearing the characteristics he seeks. Darwin needed to find the agent, or mechanism, that would account for the changes he had observed in nonhuman nature.

Then, in 1838, two years home from his voyage, he read Thomas Malthus.

Like many intellectuals of his day, Malthus was a member of the clergy. (This would likely have been Darwin's fate had his father been less wealthy.) Malthus preached from the pulpit on Sundays, received a stipend, enjoyed a respected social position, and was otherwise free to pursue other interests, chief among them being the emergent discipline of economics or, as he put it, "political economy." Preacher he may have been, but Malthus was a dry-eyed fellow who never troubled to put harsh facts in a happy frame. It is fair to say that he, as much as anyone, put the "dismal" in the "dismal science."

In 1798 he published *An Essay on the Principle of Population*, in which he postulated that when a society's food supply increased, an increase in population, powered by better nourishment, would follow. Abundance would soon become scarcity, however, because a greater number of mouths would consume the surplus, and society would revert to the previous, barely adequate *per capita* availability of food. Malthus drew the conclusion that, aside from brief periods of surplus, human existence among the poorer classes would remain a harsh struggle for existence. He saw no possibility of achieving permanent plenty for all.

In Malthus's view, Utopia lay forever out of reach. With people seeking more food than might be supplied, competition was guaranteed to be perpetual, and the strong and disciplined would inevitably dominate the weak and improvident.

Darwin's reading of Malthus inspired him to view plant and animal populations similarly. A grassy meadow produced more seeds than might ever grow into mature plants, yet certain grasses prevailed and others did not. The same applied to mice, fish, and Galapagos finches. The creatures that competed well—by finding enough to eat and avoiding being eaten—would grow to maturity and have the best prospects for producing offspring of their own. If a bill of a certain shape endowed a finch with the ability to forage better than its cousins whose bills were different, that characteristic would be transmitted to the next generation because the best-endowed finch would likely have the greatest reproductive success. In this fashion, a variety of small, incremental changes well suited to the challenge of existence might be magnified over time, potentially spreading through a population and transforming it.

Now Darwin had his mechanism. He called it "natural selection."

He would use the phrase in the title of his magnum opus: *On the Origin of Species by Means of Natural Selection, or The Preservation of Favoured Races in the Struggle for Life.*

33

Saldang Clinic

LET US now praise feet and all their parts—the curve of the arch, the stoic heel, the pliant toes. To bear witness to feet is to appreciate the sublimity of evolution, for the foot is the result as well as the agent of our election to stand and walk. And to walk and walk. The footprint, laid down in savanna dust, was our first signature on the surface of Earth, and in time we pressed it into Arctic snow and tropic mud, on mountain, forest, desert, and steppe, from ocean to ocean, and even on the moon.

The feet of Saldang now march to our clinic. We have set up within the grounds of the village school, where tuition is fifty kilos (110 pounds) of dry dung for the year. The school is the work of Western humanitarians, and the good they have done is real. Because of a similar school our fellow clinician Wangmo studies nursing in Kathmandu instead of herding goats on barren hills.

Again we rinse, soap, and scrub. Again we encounter the gnarled toenails, the pungent odors, and the tongues stuck out in shy hilarity. We are a merry group. Our clients chatter freely and incomprehensibly to us, and the feeling is one of casual, if surreal, friendship. Pete and I are mostly silent. The work requires concentration and the waiting line is long.

Of Saldang, Peter Matthiessen wrote,

> In such barrenness, the neat aspect of houses, walls, and fields speaks for the strong spirit of these villagers, who constructed spears to drive off bandits, and can dance so merrily when their food is almost gone . . . One day human beings will despair of grinding out subsistence on high cold plateaus, and the last of an old Tibetan culture will blow away among the stones and ruins.

Matthiessen's observations are decades old, yet Saldang still clings to its steep, bare slopes. No bounty or excess cushions life here. The dogs

I have seen in Upper Dolpo I can count on one hand. I have heard one rooster and spotted two cats in three weeks. There are no pigs, which cannot tolerate the cold, nor any other animal that might subsist on what humans discard. Surplus does not exist. Buckwheat, barley, mustard, potatoes, and little else will grow at this altitude, where summer prepares to depart as soon as it arrives.

A single willow survives in a corner of the school grounds. It is only a whip of a tree, but for weeks I have seen no plant half as tall. The children go to it when classes are done. They like to taste the sap, which is sweet. Perhaps it is the sweetest thing in their world. Bees, honey, and all but a few beleaguered fruit trees are impossible in this attic of the planet. Perhaps the willow sap also confers a mild analgesia. Salicin, found in willow bark, is a precursor chemical of salicylic acid, which in the laboratory evolved into aspirin. Alas, there is precious little aspirin in Dolpo, or any other painkiller, save for barley beer, the local moonshine called *rokshi*, and the rare, coveted bottle of Chinese vodka.

One might expect the people living under such grim circumstances to be sour, but we have met only courtesy and willing smiles. Adults, of course, wear many masks, but not children. Even the scruffiest among them possesses a cheerful, impish vivacity. One small mob of tykes caught sight of Pete, Becky, and me strolling down a lane. In a frenzy they shouted "namaste" again and again, across two hundred meters of empty field. Their joyful yelling continued for as long as we remained in view.

Our camp sprawls amid the ruins of good intentions. I hung clothes to dry on the lower struts of a defunct cell-phone tower, the relic of a failed attempt to bring modern communications to rural Nepal. Near it stands a windmill for generating electricity, built by a Western NGO, but the generator is inoperable, Saldang's talcum-fine dust having clogged its bearings. In the same complex as the lovely school, we encountered the locked doors of a medical clinic whose resident nurse appears to have gone on extended holiday. No one can say if she is expected to return.

Our clinic briefly fills a void, but we cannot know if the difference we make will last. The care we deliver is mainly palliative, the relief temporary. The toothbrushes we hand out have value, as do the Little Suns, corrective lenses, and sunglasses, the last of which we gave away today. Occasionally we make someone well, or even save a life, as may have been the case with the young woman who gave birth days ago in Tinje.

Harder to measure is inspiration. All the clinicians, Nepali and American alike, feel that the work they do under these wind-whipped conditions shifts their perspective, and this is also true for supernumeraries like myself: we bear witness, we try to act with compassion, we don't know what consequences may ensue, but our not knowing would make a poor justification for not acting.

And then comes surprise, which attended the case of nine-year-old Pema Nurbu. His parents brought him to the clinic because his headaches never ceased. His eyes also twitched strangely—a nystagmus, in medical terms. Michael Lobatz, the neurologist, feared a brain tumor. Long conferences among the parents, Lobatz, Roshi, Sonam, and others followed. The boy and his father would travel to Kathmandu. Roshi would guarantee the costs. Sonam would make arrangements. The boy would have a brain scan. Lobatz was prepared, if surgery were called for, to link the family to expert colleagues in India.

No one could have foreseen the outcome. Happily, no tumor was detected, and the nystagmus proved to be congenital. The scan, however, revealed a severe sinus infection, clearly the cause of the headaches, which subsequently eased under treatment by Dr. Sonam. The boy stayed in Kathmandu. He had been a poor student in Saldang, unable to study, but now, relieved of headache, he excelled. The following year the Nomads Clinic returned to Dolpo but got no nearer Saldang than the Dho Tarap valley. No matter. Pema Nurbu's parents walked all the way from Saldang to Dho, a round trip of at least eighty kilometers. They wanted to see Lobatz, Roshi, and the others. Their purpose was not complicated. It was simply to say thank you.

34

Doing the Math

AT FIRST light our muleteers scurry across the slopes to gather their drowsy and starving charges. They whistle, shout, and hurl well-aimed stones. I watched one man pound a pair of steel spikes into the ground and run a tether between them, to which he tied each mule by a fore-foot. Six or seven mules to a line, then he set another. The muleteers feed their animals handfuls of grain as they strap on the packsaddles. The lucky mules get actual nosebags with perhaps a pint or more of oats. Each muleman tends his animals in his own way. For the better part of the night the mules ate greedily, nipping millimeters of already shaved stubble from the prostrate gorse that covers the hills, but they could not fill themselves. The calories thus gained can scarcely have exceeded the calories they expended to get them.

Our record keepers report that we've now treated 823 patients, com-pared with seven hundred for the entire expedition last year. Rigdzin even found the missing medical bag, lost these three weeks within a larger duffel. The antibiotics it contained have restored the powers of the expedition's Western clinicians, and everyone, in spite of a few viruses, lingering respiratory infections, and general weariness, is fit to travel. But we've made a change of plan.

The drizzle that pattered on our tents last evening turned to snow in the night. In the high country to the north and west, our intended route to Bhijer and its fabled monastery weaves across a maze of interlocking ridges. A white blanket now covers those ridges, rendering our path invisible. Worse, clouds in that direction have snagged on the heights. Somewhere in the mist stands Neng La, a 17,600-foot pass that we must cross to reach Bhijer. Even in favorable conditions, finding the right trail to ascend the pass might take some doing, and attempting the pass in a whiteout would invite calamity. Prem, Tenzin, and the senior guides judge the risks too great. We will not go to Bhijer. Instead, we will turn southwest, directly toward Shey. Along the way we will visit another ancient monastery called Namgung.

Setting out, we pass a wall of mani stones the size of a small moraine, after which the trail steepens toward a crest of ridge and an egg-white sky. On every knoll and jutting cliff a chorten stands sentinel. Although a far horizon shows scraps of blue, for now we walk under a somber, muffled vault, the trail a furrow that centuries of hooves and human feet have scratched into crumbling sandstone.

The land falls away into invisible valleys, so that the clouds that fill them lie almost at our feet. The omnipresent choughs swirl as often below as above us. They sing, *pwow, pwow-wow*, like taut bowstrings plucked by the wind. Amid them, I feel almost airborne, floating through the bottom of the sky.

The fast walkers have gone ahead; the slow ones are behind. I am in my middling position in the expedition's mile-long line, with only the ghostly aunts and uncles for company. My thoughts wander. An uncle whispers in my ear. It is Herman Melville. This country of high ridges calls to mind his Catskill eagle, which can "dive down into the blackest gorges, and soar out of them again and become invisible in the sunny spaces."

Melville's black gorges stood for what some of us might today call depression but which he believed to be an honest, if dark, view of existence, one that acknowledged, as Solomon did in Ecclesiastes, that "All is vanity." In the end, nothing can last. And so nothing matters. The final destination is always oblivion. For Melville this was an ultimate truth. Life was purposeless when you saw it clearly, and a parade of illusions when you didn't. He called Ecclesiastes "the truest of all books" for it was built from "the fine hammered steel of woe." To dwell in such meaning, however, was dangerous. Melville well knew that Solomon's dark truth could overthrow the psyche: "There is a wisdom that is woe; but there is a woe that is madness." How to survive the tractor-pull of despair? Melville's answer was to aspire to the resilience of the Catskill eagle, the noble bird that not only dived into the gorge of woe but also soared out of it.

Today's environmental dystopia presents a gorge of woe if ever one existed. Melville's constants of the human condition—venality, brutality, ignorance, and a thousand other sorrows—have not changed, and now, in addition, our planet simmers toward a boil. If conjugating the probabilities of climate change, from drowning coasts to the spread of heat-driven epidemics, does not impart a twinge of dread, the fate of the oceans might grab one's attention: 90 percent of large fish gone since 1950, the Great Barrier Reef and every other major coral system on life support, the dead zone off the mouth of the Mississippi swelling like a tumor, a sprawl of plastic garbage the size of Texas filling the gyre of the northern Pacific, toxic algal blooms poisoning even the Bering Sea and its rookeries of raucous seabirds. I won't go on. Sure, there are feel-good, local stories that counter the bad news, but the macro-trends are stark. To contemplate them in full, not shying away, is to feel yourself drawn down a rabbit hole of hopelessness. This is the reality of Petri Earth. Uncle Herman never glimpsed the worst, save for the slaying of the whales.

A few years ago I joined an expedition in central Laos. Two days by foot and pirogue took us to the last villages. Then we plunged into the forest beyond. First there was incomparable beauty. Gibbons singing in the morning mist. The merry racket of laughing thrushes. Then increasing carnage. Under forest canopies tall and thick enough to hide the sun, we encountered hedgerows of chopped brush stretching kilometers along the length of a ridgeline or rim-to-rim across a canyon. Every few meters, a gap was left in the hedge, and in every gap lay a loop of wire. A snare. No creature could move through the forest without meeting a hedge, and none but the largest could leap across it. Most wildlife had to pass through the gaps, and the snares were sensitive to the lightest touch. I saw a silver pheasant, a bird the color of moonlight, dangling from a snare. Its feathery step had dislodged the trigger of the snare. The trigger caused a bent-over sapling to snap back. A wire tied to the sapling shrank its noose around the pheasant's foot and flipped the bird into the air. Upside down, the pheasant flapped frantically awhile, then twitched, then finally hung still until, days later, it died.

Delicate the snares may have been, but they were also strong enough to hold a wild pig or a full-grown deer. The trapped animal might pull against the snare until its strength was gone and not escape. Or it might tear off part of its leg in the attempt. We saw that too.

The makers of the traps tended them infrequently, leaving their victims to suffer and rot. In a few days we destroyed almost nine hundred snares, collecting and carrying away their lethal steel wire (bicycle brake cable, ubiquitously available). In the snares we found the rank carcasses of birds and badgers, mongooses, monkeys, and muntjacs (Asian "barking" deer). Driving this remorseless "defaunation" of the forest, which is epidemic in Southeast Asia but hardly confined there, is a hunger for the remedies of Traditional Chinese Medicine and its variants.

Poor villagers from Laos, Vietnam, and neighboring countries do the capturing and killing, but the ultimate consumers include the newly wealthy of East Asia's expanding economies. The demand for monkey hands, turtle blood, python fat, bear gall, and powdered antler, not to mention the flesh of porcupines and other animals for restaurant fare (anything caught wild is considered a "health" food), has reached new heights. Such items also confer status: in certain circles it is considered cool to flaunt your ability to buy them. Many people know that Asia's hunger for folk remedies has pushed rhinos to the brink of extinction and extirpated tigers from all but fragments of their habitat, but scores of less charismatic species are also vanishing. Such is the war against nature in the name of human health.

The carnage I witnessed posed a paradox: the forest was tragically empty, yet it was also full of beauty, its vegetation so vigorous I thought I could almost hear it growing. Vines laced through the trees like the rigging of a tall sailing ship. The rivers were limpid and musical, the birdsong lilting and constant. Even eviscerated, the forest harbored an ethereal enchantment.

The potential of the forest to generate life, if given the chance, can never be doubted. Of the Galapagos, Darwin said, "One is astonished at the amount of creative force, if such an expression may be used, displayed on these small, barren, and rocky islands." The same is true of a forest in the tropics, but times ten. Or ten thousand.

I came home heartsick from the slaughter I had seen. One image particularly lingered: a red-shanked douc, perhaps the world's handsomest monkey, dangling upside-down from its snare pole, its fur still soft and luminous, the earth only inches past the reach of its delicate hands. I felt haunted by its slow death from dehydration and despair.

Some months later I visited Canyon de Chelly in the Arizona portion of the Navajo Nation. If a place can be more overgrazed than the bald slopes of Saldang, it is Canyon de Chelly. From an overlook above the main canyon I spied a white horse midway up a dune at the bottom of a cliff. The dune was the same red as the sandstone above it, and it was corrugated with sheep trails. Thorny, inedible green shrubs stippled the sands, where nothing else appeared to grow. I could not imagine what the horse was finding to eat. The land was battered, almost barren. And yet I could not take my eyes from the view: a bright white horse amid emerald shrubs on rust-red sand. The color, the contrast, the unadorned solitude of the horse—all the elements of the tableau converged in a way that was more than pretty, more than merely picturesque. I felt stunned, and then inspired, although I could not say how. I snapped a dozen photographs, sensing that the horse and the chromatic land on which it grazed answered a question I had not known how to ask.

Not the first time I looked at the photographs, but much later, I realized they expressed something I had failed to grasp in the Lao forests.

The realization was almost mathematical. It said that Earth's beauty is inexhaustible. Even where the world is most diminished, beauty remains. The forces that erode the life of the planet can reduce but not eliminate that beauty, for beauty is intrinsic to the planet. Or if not to the planet, then to the way we Sapiens have evolved to see it. And the beauty belongs to us, inheres in us, and needs to be conserved in us too, for we are a part of the planet.

That was not all: if beauty is infinite, then the need—no, the obligation—to defend such beauty is also infinite. It will last as long as beauty lasts, and so the obligation will have no end. The image of the white horse on the abused dune gave answer to the wisdom that is woe. It said that defending the beauty of the world was a calling that would never go silent. And it said that to serve such a calling would produce something that all of us seek, which is *meaning*, durable and real. The moment of understanding felt buoyant. It was as though I'd caught a gust of uplift blowing from a deep, black gorge that carried me suddenly upward, into the sunny spaces of the sky.

35

Survival of the Sexiest

BEAUTY POSED a problem for Darwin, which he addressed with his usual thoroughness. Natural selection, he realized, failed to explain marvels like the resplendence of the peacock's tail. It explained predator-prey relationships, tests of strength and dominance, and other adaptations that helped organisms survive in new or changed environments, these being the tensions swept up in the phrase "survival of the fittest" (which Darwin did not coin, although he accepted it). Natural selection explained much, but it did not explain everything, least of all beauty.

The ethereal song of the hermit thrush. The elaborate courtship structures of the bowerbird. The beard of the bearded monkey. Some evolutionists, more "Darwinian" than Darwin, have pretzeled their logic to argue that all such phenomena manifest some aspect of fitness, however cryptically, and that they guide breeding adults (usually the females) in selecting mates whose offspring will have superior prospects for survival. Darwin considered such thinking to be an overextension of his theory. He did not agree that the size and splendor of the tail of a particular peacock meant that its progeny would be hardier than those descended from peacocks with less glorious appendages. Instead he thought it probable that peahens had evolved simply to prefer males with big, splendid tails. The peahen's choice certainly included some kind of awareness of her prospective mate's health and vigor, but, all things being equal, her selection came down to a matter of taste, which Darwin ascribed to the peahen's notion of beauty. Moreover, he marveled that, as far as the tails of peacocks were concerned, peahens and humans largely agreed on what was beautiful and what was not. He concluded that peafowl—and many other species—possessed an aesthetic sense similar to ours. This judgment undergirded his view that natural selection could not account for all of nature's variability. Instead, contingency, randomness, and caprice, driven by what he called sexual selection, took over where the grim realities of natural selection left off.

Darwin had meant to address this subject in *On the Origin of Species*, but *Origin* outgrew his original conception, risking unwieldiness, and he recognized that a decent treatment of sexual selection, besides greatly enlarging the book, would further postpone its long-delayed completion. And so he put off the contemplated work, but the subject hardly lay dormant. It inspired him to gather volumes of evidence, and ultimately he devoted a second magnum opus to it, comparable to *Origin* in length, heft, and ambition. This was *The Descent of Man, and Selection in Relation to Sex*, which appeared in 1871, a dozen years after *Origin*.

It is easy to view Darwin as a scientific recluse, toiling obsessively over the intricacies of the natural world while immured in his estate at Down, Kent. He was, after all, kept at home by delicate health. But he was hardly detached from the affairs of the world. The title of Darwin's second great theoretical work announces implicitly his intention to address one of the most contentious issues of his age: the legacy of slavery and the relations among the races of humankind. Both Darwin and his wife, Emma (née Wedgwood, of the pottery dynasty), hailed from families that had cheered Britain's abolition of the slave trade in 1808. Darwin was aboard the *Beagle* in 1833 when Britain finally abolished the actual practice of slavery, which would take the United States decades more and a civil war to accomplish. He was well aware that many apologists for slavery believed that Black Africans, and for that matter Native Americans, originated independently from the white race and that their separate genesis accounted for their presumed inferiority and therefore justified the horrors of both enslavement and extermination. Darwin accepted none of this, and while in Brazil, an argument between him and Captain FitzRoy—precipitated by FitzRoy's statement that slaves were better off in the care of white masters—caused a breach between the two that nearly ended Darwin's voyage.

Darwin approached the problem of human origins as he approached other questions—empirically, studying variations among humans much as he did among birds. He asked why the Native Americans and Black slaves of Brazil should differ so greatly in appearance when their places of origin on different continents shared a similar tropical environment. As the *Beagle* completed its circumnavigation of the globe, he investigated similar questions in Polynesia, New Zealand, Australia, and Mauritius. Ultimately he concluded that variations in curliness of hair, shape of nose, thickness of lips, and other traits producing observable differences among human populations had arisen for much the same reason as the extravagance of the peacock's tail—because the members of those populations liked how those features looked. They preferred them, were attracted by them, and they more often selected mates who possessed them than mates who did not. Most assuredly, notions of beauty were not the only factors driving selection, but they were in the mix, and Darwin essentially concluded that the rainbow of contemporary humanity looks the way it does because the ancestors of the various races, going back into the deep anthropological past, developed different ideas about what was beautiful in the human face and body.

Selecting a mate was never the only, or even the most important, arena for the decisions of ancestral humans. There were myriad questions to answer: is this food ripe, does this place look safe, does it seem to have the things we need? An aesthetic awareness would seem consistent with a developed sense of the suitability of the world. Good things look good, bad things ugly. An aesthetic criterion can migrate from the necessary to the useful, and thence to the pleasing, and from there to the complexities of still more arbitrary judgments. Like many other animals, humans cannot live without a sense of what looks right. Prettiness, in a simple reading, arises from a happy union of form, composition, and color. But beauty takes pretty a step further, at least for us modern Sapiens. Beauty is more than merely pleasurable. Beauty becomes beauty when prettiness is joined with wonder. The world and the things in it often inspire awe: we stop in our tracks, we say, Oh my goodness, would you look at that? When the awe or startle is agreeable—and sometimes, as with a storm-tossed sea or forbidding mountain top, it is agreeable to be scared or shocked—the scene before us, or the person, or whatever we happen to be looking at becomes beautiful.

This innate aesthetic also transfers to ideas. The pretty fit of coherent data becomes beautiful when its implications inspire awe and wonder. The theory of evolution, by this accounting, is steeped in beauty, as is the fit of creatures into their habitats. So too is the beautiful fit of the continents and the theory of plate tectonics. As sublime as any "fit" on Earth is the double-helix structure of DNA.

In the early 1950s, multiple laboratories raced to unlock the secrets of the code-bearing molecule of life. At the University of Cambridge, two microbiologists, James Watson and Francis Crick, built a model for the structure of DNA that they thought explained how the molecule actually functioned. They feared, however, that their close collaborator, the pioneering X-ray crystallographer Rosalind Franklin, would reject the model as too speculative or just plain wrong. Their relations with "Rosy" had grown testy, and she was a resolute and remorseless critic. They were greatly relieved, therefore, that when she saw the model, she accepted it instantly. Watson later wrote, "Like almost everyone else she saw the appeal . . . and accepted the fact that the structure was too pretty not to be true."

Different tellings of the story have attributed the coinage of the final phrase to different members of the trio, and sometimes the word "pretty" is replaced by "beautiful." That is as it should be. The double-helix structure divined by Watson, Crick, and, yes, Franklin (although she does not always get the credit she is due) is wondrous indeed. It is too beautiful not to be true.

Earth brims with that microscopic architecture. Strands of DNA abound on land and sea, even within rocks deep underground and on particles of dust wafting through the air. Its variations are as numerous as the stars. If beauty inheres in such a marvel, then surely nothing in the universe is more beautiful than the frail blue dot on which we ride through the vastness of space.

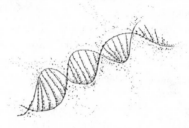

36

The Cranes of Namgung

WE ARE in the mess tent, in camp on a ridge towering above the few houses and stolid gompa of Namgung. It is afternoon, and the kitchen crew have set out thermoses of sweet tea and snacks of pistachios and crackers. We are quite spoiled by this treatment but also grateful for it after a long day's walk. A few hours ago and a thousand feet lower, we visited Namgung's blockhouse gompa, which stands amid an alley of chortens and a virtual talus of mani stones. Below it, several narrow barley terraces cling to the canyon's lower slopes. Namgung is a dark place, so deep does it lie in its chasm, and its gompa is also dark. Only in the beams of our headlamps could we appreciate the intricate, mazy designs that spread in primary colors across walls and ceiling. They rendered the gompa's interior covertly joyful.

Now, in our camp on the high ridge, we are awash in the slanted light and long shadows of the day's late sun. This will be the highest campsite of our journey—15,234 feet. It is also perhaps the most beautiful, for the slopes around us have turned crimson with the autumn foliage of cinquefoil and mountain sorrel. From here, the views northeast into Tibet embrace a receding sea of stone-gray peaks that merge into gray-bottomed clouds.

The crowded tent is redolent of tea and cinnamon. Suddenly a hubbub stirs outside. "You better come," someone shouts. The tent empties. Outside, everyone stares into the sky. "Look up!" "Oh my god." "Birds." "Geese," someone offers. "No," I say. "Cranes!"

Hundreds of them. They labor high above us in ragged lines, wave after wave, an entire sky in motion, their straggling formations shifting, dissolving, reforming. The birds are easily half a mile overhead, no more than specks, but we can see the long extension of their necks and the longer span of their powerful wings, which beat to a single rhythm, a thousand earnest metronomes. Someone—it may have been me—asks for silence. And our exclamations quickly stop. It takes a moment to recalibrate to the quiet, but then we hear the birdtalk spill down to us, the creaking, croaking positional calls, long-throated and eager. Now we eavesdrop on a foreign world. We hear the cranes' guttural chatter, which they first voiced in the Miocene, a language millions of years more ancient than humanity.

The birds are almost certainly demoiselle cranes (*Anthropoides virgo*), the smallest of the fifteen extant species of crane. Of all the fifteen species, they are also the most dauntless. Certain populations undertake a migration as arduous as any on the planet. They winter on the plains of the Ganges and summer on the Mongolian steppe, with the colossal obstacle of the Himalaya in between. There is evidence that some demoiselles flank the Himalaya to the west on their northbound journey in spring, returning in a great circular arc over the Himalaya in the fall. Other populations labor up and over the mountains on both legs of their migration. They have been filmed in flight as high as twenty-six thousand feet, in frigid temperatures amid vicious winds. They brave such altitudes by night as well as day. No one knows how old their migration may be, but certainly its origins lie deep in the Pleistocene, or earlier, which raises the possibility that its route may have been engraved in their neurons before the spectacular late uplift of the Himalaya began, and that as the mountains rose, the birds obligingly soared higher and then still higher.

Their ingrained, collective fearlessness is the more inspiring for being embodied in a form so elegant and seemingly fragile. Like all cranes, demoiselles are sleek and graceful, with extravagant breast and tail plumes that project foppishly outward, past the contours of the body. A stripe of white above the eye bends behind the dark neck, and ends in a delicate, wispy dangle, like a burst of hair above a dandy's collar. The fierceness of their red eyes is striking, and so is the adaptive fitness of their short-toed feet, which are different from the feet of other cranes and well-formed for running on the grassy steppes of their Mongolian summer home. Beautiful are their calls, many and varied, and beautiful are their dances, which, once begun, will spread contagiously through a flock until every bird is leaping and flapping, puncturing the air with a darting bill.

Some say the first lines of poetry written in India, which may have been the first formal lines of poetry in the world, describe a crane mourning its mate. The crane populates the imagery of certain Asian religions as prominently as the ibis inhabits the mythology of the pharaohs. We have named our tallest machines after cranes, and when we stare into the sky to wonder at the spectacle of their flight, we crane our necks. The beauty that one crane recognizes in another is perhaps not different from the beauty with which cranes entrance us Sapiens. Without restraint, they give their beauty to the world.

Minutes after the cranes pass, an orange-red sunset suffuses the sky with a light both lurid and tender.

37

Flowing Mountains

BEFORE CLIMBING to camp, I follow several friends to a bend in the Namgung canyon. They lead me to an outcrop of hard, water-polished limestone. We splash the rock from our canteens to bring out its hues of maroon and gunmetal blue. Encased in the outcrop, framed by white streaks running through the rock, is the fossil of an ancient sea creature, a nautiloid. The name of one of its descendants, the ammonite, gives a hint of its appearance. Ammon was a deity in ancient Egypt who wore a headdress of ram's horns. The spiral of the ram's horns and the curl of the ammonite are the same. Although ammonites are long extinct, their shape lives on in nautiloids that survive today, including the elegant chambered nautilus, whose shell is coveted for its delicate, sculptural beauty.

I cannot guess the age of the fossil at Namgung, but hundreds of millions of years ago, such creatures were a top predator of the world ocean. Today their line continues in squids, octopi, and their cousins, some of which (squids) internalized their ancestral shell, while others (octopi) dispensed with it altogether. In the fossil before us, the shell has turned to rock, its spiral precise, its sections blue-black and distinct, progressively enlarging as the shell spiraled outward to accommodate the growing creature inside.

Namgung now stands two and three-quarters miles above the vanished sea in which the living nautiloid swam. In death its body mingled with sediments that buried it deeper and deeper, even as the level of the original seabed subsided over eons of time. As heat and pressure increased, the nautiloid's remains were forged into stone. And then, for eons more, the movement that buried it reversed, and the former seabed rose. Toward the end of the fossil's elevation, water and ice cut away its lithic covering to expose it to the air, where we could see it and where it will remain for no more than a geological instant. Soon enough, it will be ground into sand, the grains of which will commence a return journey toward another sea, where one day they may bury the shell of something else.

In the thirteenth century, when the Crusades ravaged the Holy Land and Genghis Khan, at the head of a mounted army, swept into Europe, the Zen master Dogen wrote, "When you take one view you see mountains flowing, and when you take another view, mountains are not flowing." His immediate concern was the subjectivity of perception, not a process of geology, but his observation applies to both. Dogen's *Mountains and Rivers Sutra* would tell us that all things are impermanent, both within the universe we construct from our senses and within the physical world that is subject to time.

From camp at sunrise the next morning the view northeast toward Tibet attests to that reality. The crimson sorrel and cinquefoil foretell the onset of winter, and the distant mountains spread sealike to infinity. The impression is magnified later in the day when we summit the 16,700-foot pass of Sé La, en route to Shey. From that height the horizon falls back even farther, and still the sea of mountains appears limitless. (It was this view that inspired George Schaller, when he glimpsed it upon leaving Shey in 1973, "to travel on and on into the unknown" of Tibet. He has since made more than thirty-six trips there.)

Most astonishing in the view from the pass, every mountain in the sea of peaks appears to be roughly the same height as every other, conferring an unexpected uniformity to the vista, as though the mountaintops were indeed not solid but were the crests of waves, and the waves seem pregnant with motion, as though they could flow or, at the scale of Earth-time, are flowing, their beauty impermanent and doomed, yet belonging to a cycle in which new beauty must ceaselessly arise.

38

Flowing Seafloor

FRED VINE was a graduate student when he glimpsed the literal flow, not of mountains, but of their opposite, the seafloor. He had embarked on a doctorate in marine geophysics at Cambridge University and was assigned to a newly minted PhD, Drummond Matthews, as a researcher. Matthews was only a few years older than Vine and knew nothing of Vine's assignment to him, for at the time he was aboard a vessel in the northwest Indian Ocean, mapping an undersea formation called the Carlsberg Ridge. The year was 1962.

Vine, a Londoner, fit the caricature of the intellectually audacious graduate student. He gave the impression of a large brain tottering on a scarecrow frame, his brow wide and high, his jaw narrow, his eyes as dark as his skinny necktie. While still an undergraduate he had devoured everything written by Harry Hess, the former naval commander who had sounded the depths of the Pacific during the war. In 1962 Hess came to Cambridge to give a talk on "The Evolution of the North Atlantic," in which he described the restless geology of the Mid-Atlantic Ridge. If he used visual aids, he no doubt showed a version of Marie Tharp's map. For Vine, the lecture was an inspiration. His ideas took flight.

It was exceptional good luck that he was assigned to work with Matthews. Their closeness in age helped make their collaboration equitable, and Matthews was already far along in accepting the idea that formation of seafloor was an ongoing, continuous process. He had concluded, for instance, that ocean ridges like the Carlsberg formation (named for the brewery, which funded the expedition that identified it) resulted from the eruption of volcanic material through fissures in Earth's crust. The rent that formed the Carlsberg Ridge was an extension of the trough of the Red Sea and the Gulf of Aden, which were known to be seismically active.

Matthews and Vine had their offices some distance from the main Cambridge campus, in a coachman's quarters above a former stable where seafloor cores were stored in the stalls. Vine's job was to analyze the magnetic data Matthews brought back from surveys of the Carlsberg Ridge. It was understood that rock formations both on land and under the sea varied in their magnetic orientation. These variations were generally thought to derive from the kind of rock involved and from forces, like weathering, that had affected the rock. Efforts to map the variations, however, were in their infancy, and Vine set to assembling the Carlsberg data in two- and three-dimensional models. To do this, he drew upon a new tool just then coming into use, the digital computer, and he borrowed software from other researchers and tinkered with the programs to improve the accuracy of his models.

Part of the challenge was to fit Carlsberg's magnetic data to three-dimensional topographic descriptions of the seafloor. This sort of modeling had been attempted for other formations prior to Vine, but no one had done it well, not least because the data were usually poor, the transects being disparate and imprecisely correlated. Working with Matthews's superb field records, however, Vine was able to construct a detailed picture of magnetic orientations across the entirety of the Carlsberg Ridge.

The map that resulted was striped like a zebra. Some of the bands of seafloor were oriented toward a north that was more or less where north is today. But alternating bands showed "north" pointing southward, as though the electromagnetic poles of the planet had flipped. The possibility of polarity reversals was not a completely new idea—others had suggested it—but most geophysicists viewed the notion with skepticism. Vine's work offered new, strong evidence that through the ages Earth's magnetic polarity had repeatedly reversed itself from north to south and back again.

Then Vine drew a further conclusion: the bands of alternating polarity in the Carlsberg Ridge had not been induced by forces acting on the rock after it was formed; instead, they were *remanent*—they remained in the rock from the time of its formation. He hypothesized that, in cooling, the magma that erupted through the seafloor oriented itself to Earth's polarity as it then existed. This orientation was preserved as the magma hardened into basalt, and it persisted in the rock unchanged through the ensuing eons. Because the formation of seafloor was roughly continuous, the rocks comprising it contained an unbroken record of Earth's magnetic orientation, age after age.

Vine connected three emergent concepts—seafloor spreading, polarity reversals, and remanent magnetism—in the draft of a paper in which he argued (1) that as seafloor formed through time, it gradually spread outward from fissures like the Carlsberg Ridge and (2) that remanent magnetism could be used to deduce the sequence of polarity reversals from the oldest sample of a given transect to the youngest. That's a mouthful, so here it is put another way: seafloor material erupts at mid-ocean ridges more or less continuously, and the eruptions record the polarity of Earth as it existed when the erupted material cooled. The bands of polarity that are farthest from the fissure at the center of the ridge are the oldest—they have spread the most; those nearest the center are youngest. The alternating bands of normal and reversed polarity provide a record of spreading through time.

Harry Hess had conceived of seafloor spreading as a conveyor belt. Fred Vine said it was more than that: it was also a recording device.

Drum Matthews was away on honeymoon when Vine first sketched out his ideas in writing, and as he had no one else to talk to, he broached his theory to Maurice Hill, the redoubtable head of his department. Vine later recalled, "[Hill] thought I was totally mad . . . he just looked at me and went on to talk about something else." Vine presented his theory to another senior colleague who murmured encouraging words but quickly said "no way" when Vine invited him to coauthor a paper on the subject.

Upon his return, Drum Matthews took a different view. He saw merit in Vine's concepts and worked with him to strengthen the paper with additional field data drawn from Carlsberg Ridge. The result was "Magnetic Anomalies over Ocean Ridges," which appeared in the prestigious journal *Nature*, bearing both their names, on September 7, 1963.

Alas for Vine and Matthews, the paper "went over like a lead balloon." Almost no one paid attention.

Coincidentally, another paper expressing the same synthesis made the rounds of *Nature* and the *Journal of Geophysical Research* that year and was rejected by both. Its author, Lawrence W. Morley, headed the Geophysics Division of the Geological Survey of Canada. He independently came to conclusions almost identical to those of Vine and Matthews, but was unable to find a forum for them until he presented them orally at a scientific meeting in Quebec in June 1964. He was Alfred Russel Wallace to Vine and Matthews's Darwin. Ultimately their shared conception of seafloor formation would be known as the

Vine-Matthews-Morley hypothesis. But a hypothesis was all it was. Until somebody pegged their putative record of polarity reversals to specific geologic periods, and until they fleshed out their model with the kind of detail that would answer more questions than it raised, the world would continue to ignore them.

Zebra-striped seafloor showing bands of alternating polarity in the seafloor offshore of Vancouver Island, Canada. After AD Raff and RG Mason (1961), "Magnetic Survey off the West Coast of North America . . . ," *Bulletin of the Geological Society of America*, 72: 1268. Used with permission.

39

Tuzo Wilson

THE VINE-Matthews-Morley hypothesis eventually took a central place in the conceptualization of plate tectonics theory, but it was not the only line of evidence building toward a breakthrough. Other researchers had obtained surprising dates from seafloor cores—all seafloor rocks proved to be comparatively young. Unlike continental rocks, none was much older than a hundred million years. Researchers were also defining new seismic zones in the ocean's deepest trenches, and recording clues about the movement of magma below Earth's crust. Virtually all the new insights lent weight to the arguments of *mobilists*, as the advocates of a new and as yet undefined version of continental drift had begun to be called. One arena of investigation exploited paleomagnetics by reconstructing not pole reversals but the former positions of continents.

The magnetic alignment recorded in the formation of rocks doesn't curve along the spherical surface of Earth; it runs straight to the pole. This means that the *angle* of magnetic orientation "dips" below the horizon, as though arrowing through the earth. Imagine, for a moment, an orange or any sphere: the angle from its waist to its top will be forty-five degrees. From any other place on the top half of the orange the angle will be less. Thus the angle of dip will be greater in rocks formed near the equator than in those from higher latitudes. By analyzing dip, geologists learned to estimate the latitude at which a given rock formed, which was often not its modern position. They soon realized that many continental rocks must be extremely well traveled. Although it was known that the magnetic north pole had drifted around its Arctic location, the continents must have moved even more. A lot more.

These new understandings found a champion in J. Tuzo Wilson. The confident Wilson possessed a gift for synthesis. Turning to the new abundance of geophysical data, he did not hedge his conclusions, as Hess had done with the term "geopoetry." Wilson was bold enough to say, "This is how the planet works," and in 1963 he presented his conception in a sweeping article in *Scientific American* bearing the simple title "Continental Drift."

Wilson warned his readers that he was "not presenting an accepted or even a complete theory but one man's view of fragments of a subject to which many are contributing." He acknowledged that the idea of continental movement remained a minority view among geologists, but asserted that his version of drift differed from its predecessors in that it did not require "continents to be driven through the crust like ships through a frozen sea." Instead, he posited that the crust itself was in motion, driven laterally by the convectional motion of the magma beneath it:

> Across the floors of all the oceans, for a distance of 40,000 miles, there runs a continuous system of ridges . . . Most oceanographers now agree that the ridges form where convection currents rise in the earth's mantle and that the trenches are pulled down by the descent of these currents into the mantle . . . Here, then, is a mechanism, in harmony with physical theory and much geological and geophysical observation, that provides a means for disrupting and moving continents.

Wilson drew his evidence from across the globe, from Arctic islands off the coast of Siberia to coal deposits beneath frozen Antarctica. He cited his own work on oceanic islands, which hypothesized that the farther an island was from the mid-ocean ridge or hot spot that spawned it, the older its rocks should be. This proved to be the case with all the world's major islands, with the exception of Madagascar, the Seychelles, and the Falklands, all of which were demonstrably continental fragments. The available evidence, said Wilson, implied that the seafloors were in motion.

Vine and Matthews published their hypothesis a few months after Wilson's synthesis appeared in *Scientific American*. Their paper may have failed to find the audience they had hoped it would, but Harry Hess read the paper and liked it. Eventually, so did Wilson. Before long, the two senior scholars visited Vine and Matthews in Cambridge, and a close collaboration between Wilson and Vine ensued. One of Wilson's contributions addressed a troubling feature of many oceanic ridges and the bands of alternating polarity accompanying them: the ridges often ended abruptly, their continuous alignment broken. The pattern of ridges and bands appeared to shift laterally, resuming in an offset position. Wilson postulated what he called "transform faults" that ran perpendicular to the main axis of the ridge. Lateral slippage along these faults, he said, released tensions that developed as the ridges formed. Such slippage accounted for the offset fragmentation. Problem solved.

Another of Wilson's contributions was the idea that ocean ridges spread at a constant rate. As a result, the width of each band of normal or reverse polarity should be proportional to the length of time that the polarity lasted: narrow bands reflected short periods, wide bands long ones. Moreover, the bands on either side of a ridge should mirror each other with the same record of magnetic polarities changing through time.

Scanning data from the northeast Pacific, offshore of Vancouver, part of an area that the US Navy had exhaustively mapped for "submarine safety," Wilson and Vine thought they found the kind of pattern that supported these hypotheses. It lay along a formation they called the Juan de Fuca Ridge, which was bounded at its extremities by transform faults, one of which was an extension of the famous San Andreas Fault. The map of reversals showed periods of normal and reverse polarity in black and white vertical stripes. It was the old zebra pattern, which, in digital terms, looked like the kind of bar code that nowadays gets scanned at a checkout counter. Oceanic ridges from around the world yielded similar patterns. Wilson and Vine believed that the bands resulted from the steady upwelling and spread of seafloor at a rate of roughly a centimeter per year.

A tantalizing further possibility emerged: if the polarity reversals at Juan de Fuca Ridge could be found in terrestrial rocks, and if those rocks could be dated, then Wilson and Vine could peg their hypotheticals to hard geologic facts. They could build a timeline showing when different portions of the seafloor had actually formed. They would no longer be theorizing about the behavior of Earth; they would be documenting it. This was no small matter. Verifiable dates would confirm a new paradigm for how the present world had come into existence.

40

The Jaramillo Event

ONE DAY I held in my hand a rock that was the linchpin of plate tectonic theory. Or rather, I held a twin of the linchpin, a rock that was left behind.

I was in the backcountry of the Valles Caldera National Preserve, an eighty-nine-thousand-acre swath of public land an hour from my home. I faced a rhyolite bluff pocked with a dozen or so shallow cylindrical holes from which samples had been cored. In one hole, the sample had been left behind. I pulled it out. The buff-gray rhyolite was light and relatively soft, and the cylinder, the size of a small juice can, had the heft of something that would feel good to throw. Behind me gurgled the Río San Antonio, a stream too small to be called a river outside the dry Southwest, but lively and twisty and rich with trout. A bluff across the river matched the one I faced. It was pocked with cylindrical holes too.

The Valles Caldera is a young land. Slightly more than a million years ago its ancestral landscape erupted with a force a hundred times more powerful than that which blew the top off Washington's Mount St. Helens in 1980. Several hundred cubic kilometers of rock, earth, and lava exploded into the sky or flowed in molten rivers spilling outward in all directions. When the eruption exhausted itself, the hollowed-out land fell in upon itself, forming a spectacular mountain bowl—a caldera—thirteen miles across and hundreds of feet deep. But the fireworks weren't over. Smaller eruptions resumed soon (in geologic terms) after the mega-eruption and have continued virtually to the present.

The consistent, if intermittent, volcanism made the Valles Caldera an intriguing geologic puzzle, and it also meant that, latent in the rocks of the caldera, lay a record of Earth's magnetic orientation through the late Pleistocene, a record that included many of the epoch's polarity reversals.

In the early 1960s geologist Robert Smith wanted to map the caldera and deduce its volcanic history, a task for which he needed dates for the caldera's principal eruptions, which in turn required specialized laboratory assistance. Unbeknownst to Smith, another geologist,

Richard Doell, was building a chronology of Pleistocene polarity events at a laboratory he had recently established in Menlo Park, California. The lab specialized in the kind of analysis that Smith needed. Both men worked for the US Geologic Survey, and they soon met. Doell described for Smith the timeline he had built and pointed out gaps where he needed data. Smith recalls, "I saw what they were doing. It was wonderful work. So I told Dick, I said, I can give you that interval between two rhyolite domes out in New Mexico, if you'll get me dates for a bunch of domes I want."

Working in the caldera was no hardship. Its immense grassy parks, or *valles*, bounded by dark mountain forests, have furnished a backdrop for scores of movies and television westerns, as well as ad campaigns for Stetson hats and Marlboro cigarettes. The fictional home of the title character in the TV series *Longmire* was set at the caldera headquarters. *Playboy Magazine* even shot a photo spread there. Doell and a young colleague, Brent Dalrymple, collected samples from many locations within the caldera, and in an especially secluded setting—the two buttes facing each other across the Río San Antonio—they detected a previously unknown event, a brief period of normal polarity interrupting a longer span of reversal. They cored their samples in pairs, a half-dozen pairs from either side of the stream.

In the lab, they calculated that the lavas forming the outcrops had erupted about nine hundred thousand years earlier and represented an important new episode in the sequence of longer periods they had already identified. They began talking about it as the "Jaramillo event," borrowing the name of a nearby creek. But it was the wrong creek. By the time anyone understood that the rivulet between the outcrops, the intended namesake of the event, was the Río San Antonio, not the Río Jaramillo, the name "Jaramillo event" was already in currency, and so it has remained.

Even before the new event bore a name, Brent Dalrymple referred to it in a paper he gave in November 1965 at a geology meeting in St. Louis. Fred Vine, who had recently moved from Cambridge to Princeton at Harry Hess's invitation, also presented a paper at that meeting, and so did Tuzo Wilson. Chatting between sessions, Dalrymple informed Vine of the Jaramillo event and remarked on the sharpness it added to his lab's chronology of polarity reversals. Vine immediately recognized that the new timescale matched the polarities he had identified at Juan de Fuca Ridge. Moreover, it pegged his Juan de Fuca data to precise dates.

It was not exactly a eureka moment—Vine had long been convinced, not just of constant seafloor spreading, but of the global character of polarity reversals. The polarities recorded in the seafloor at Juan de Fuca and in the lavas of the Valles Caldera were embedded in rocks worldwide. The proof of this was now at hand. And more proof soon arrived from other sources, showing that virtually all ocean ridges exhibited the record of reversals displayed at Juan de Fuca.

The discovery of the Jaramillo event completed the essential case for plate tectonics. It was the keystone in the arch of a master unifying theory of how Earth actually works, and it had finally dropped into place. Wegener's "fit" of the continents was no longer controversial; it now belonged to the accepted history of deep time. Much finishing work remained to be done, but the theoretical structure was established. A scientific revolution—a paradigm shift, in the language of historian Thomas Kuhn—had been launched. Within a few years, the theory of plate tectonics would be largely fleshed out. It would take its place alongside natural selection and the quantum and relativity theories of physics as a fundamental scientific understanding. In *The Road to Jaramillo*, a history of this scientific odyssey, William Glen writes,

> More was learned in a few years about the structural evolution of the oceans by magnetic surveys than was learned over two centuries about the structural history of the continents from geologic mapping. No one had ever dreamed that oceanic geology could be so simple. The generation of new oceanic crust at spreading centers and the destruction of old in the deep-sea trenches have produced an endlessly rejuvenated seafloor, generally less than 100 million years old. That crust is magnetically banded in chronologic sequence and covers much of the earth's surface.

With this most consequential realization, an outline of the true biography of the planet began to emerge. Soon the underlying concepts of plate tectonics would be applied to the study of other planets and our own moon. That we know as much as we do is a kind of miracle, a source of wonder. The modern human discovery of how Earth actually works constitutes a rare instance of a small node of conscious life learning to understand its origin and place within the vastness of the universe. It is a blooming of self-awareness at a virtually planetary level.

41

Shey

AT LAST, Shey.

Trail-weary, we round a hump of mountain and look down a long valley where rivers converge at the foot of a flood-scoured plain. The day is clear and wind-calm, and as silent as so large a space can be. The plain shines with white boulders. On a bluff above it stands the ancient gompa of Shey, thick, squat, and festooned with flags. From so great a distance we cannot appreciate the solemn proportions of its façade or the skulls with curling horns on the rostrum before its painted doors. We see only its isolation and its blockhouse bulk. It is a brave, small fortress at the end of the world. Beyond it, across the confluence of rivers, looms the towering mass of Crystal Mountain, the heart of Dolpo and the center of a mystic geography.

Arriving at Shey excited the kind of elation I felt as a child when I glimpsed the ocean for the first time. Shey dwells in the imagination no less than in the world. It seems numinous, as though it exists beyond the frontiers of the actual. Countless travelers have labored to get here, Schaller and Matthiessen among them. They made their camp in a crude house and stockpen close by the gompa. Matthiessen stayed two and a half weeks, though in his narrative it seems longer. Schaller stayed a week or so more. While there, they roamed the enveloping slopes or surveyed the mountains with binoculars or spotting scope, scanning for blue sheep, wolves, and snow leopards. Matthiessen never saw the big cat, but Schaller glimpsed one on his return homeward, after Matthiessen had left.

Theirs is the most famous pilgrimage to Shey, at least in the West, but periodically hundreds, cumulatively thousands, of pilgrims come to Shey. Some of them make a kora of sacred Crystal Mountain, a circumambulation amounting to ten miles or more of hard walking. It is said that from the top of Crystal Mountain you can see the upturned bowl of Mount Kailash in the far distance. The faithful also attest that the two mountains are linked in bonds of spiritual power. Legend holds that the tamer of the region, Tenzin Repa, conquered demons here. He flew through the sky like a Tibetan Apollo, borne aloft by a brace of mythic snow lions. On each defeated demon he placed a giant boulder, which he marked with the footprint of one of his lions (which are not to be confused with lowly snow leopards). The count of such victories runs to the sacred number of 108, and one of them occurred near a village we sighted from the Namgung trail. The devout and the curious go there to view the footprint in the rock. Amid such resonance of time and spirit, no one casually climbs to the summit of Crystal Mountain. Tradition insists that one must earn the right to do so by first completing at least seven koras of the mountain, more than seventy miles of arduous pilgrimage.

We pitch our camp on the plain at the foot of Crystal Mountain, beside a cluster of shacks where passing traders find simple meals and shelter. Shey is a crossroads. From here at least five major trails radiate like the points of a star to Bhijer and Saldang, to valleys east and west, and notably for us, to the formidable pass of Kang La, 17,552 feet, five miles to the south, which guards the return route to Lake Phoksundo and the completion of our journey. The pass, a steep ascent into thin air and a still longer descent down the other side, will test us more than we have yet been tested. The prospect makes me glad our stay in Shey will be a long one: we will have three long nights to gather our strength. I, for one, am feeling worn. My legs, and my whole body, have acquired a heaviness that will not lift.

42

Prayer Mills

THE MOST complex machine in Shey is a cluster of water-driven wheels attached to cylinders inscribed with prayers. The cylinders revolve continuously, night and day, every day of the year. The machines that contain the wheels and cylinders are prayer mills. Matthiessen praised them for "calling on all the elements in nature to join in celebration of the One."

Their design is that of a gristmill in miniature. The mechanics may be simple, but the thinking behind them is not. The prayer mills straddle a braid of rockbound, hand-dug channels split from the stream that descends from Kang La. The stream is small at this time of year, but the naked boulders of the plain attest to repeated floods that have swept the land bare. I wonder how many generations of prayer mills have been built and rebuilt since Matthiessen was here.

The prayer mills are housed in stone cabinets, six of them, their walls mortared, the wooden beams of their roofs laid side-by-side. A flow of water spins blades on the bottom of a vertical axle, causing the axle to rotate. At the top of the axle is a prayer wheel densely inscribed with mantras, chiefly (I am told) "om mani padme hum." Because the wheel spins, so the thinking goes, the mantras are as good as chanted, and their good energy goes out into the needy world.

The mills pray together. Their analog may be found in a hydroelectric dam where turbines produce electricity, but the water harnessed here generates a different kind of light. The wheels repeat their hundred mantras thousands of times an hour, every hour without cease. While you are sleeping, eating, or scratching your head, the machines are praying, nudging the cosmos toward harmony and enlightenment. At least that is the idea.

Physicists say there are four kinds of force in the universe. There is the strong nuclear force, which fuses together protons and neutrons from their constituent quarks, and the weak nuclear force, which governs the decay of one kind of quark into another. There is also electromagnetism, which binds electrons to atomic nuclei and which, when those electrons are caused to flow as electricity, powers the greater part of civilization. The fourth force is gravity. That's it. These four forces account for the existence of matter, the shape of the universe, and the interactions of all the things in it, including the generation of consciousness. No additional force has been detected that would account for bending spoons with the kind of telepathic power Uri Geller claimed to possess, or for faith healing, angels, divine miracles, or the power of prayer, let alone the power of prayer mills. This is what science says.

But not everyone is convinced. According to Roshi Joan, the Dalai Lama has an answer for the reductionism of science. When told that physics could not account for mystic phenomena, he suggested physics might be missing something. He told his scientific interlocutor that he should "look at things more granularly."

The Dalai Lama has an impish streak. He well knows that bosons, gravitons, and the other elements of quantum physics are as granular as granularity has ever got. Possibly dark matter, once it is examined, will answer questions we hardly know how to ask. Then again, perhaps many things that seem unexplainable by physics can be explained another way.

The brain is mysterious. It mediates most of what we feel, imagine, and do (our "gut" literally and metaphorically mediates its share too). Significantly, the brain is the locus where expectations form, and expectation can shape pain, health, and much of experience. There is no consciousness without conditioning, no *tabula rasa*, not ever. We expect, and so we are. Different levels and kinds of expectation, studies show, can release hormones and other neurochemicals that alter the homeostasis of our bodies. This is called the placebo effect, and it really needs another name, for *placebo* carries a connotation of falseness, and the tangible physiological changes brought on by expectation—whether hope, dread, or apathy—are anything but false.

And so the Navajo singer or Hmong shaman or Christian faith-healer can elicit physical changes in a patient because interaction with the healer releases pharmacological powers within the patient's endocrine and neurological systems. This may be how many forms of traditional healing—at their outer limits, beyond the demonstrable efficacy of many herbal and dietary remedies—claim their victories. The unreckoned forces that science has not measured probably reside, not in the ethers of the universe, but within us, and we, alas, are the limit of what they can affect.

Neither the prayer mills of Shey nor the ineffable hunger for good that lies behind them will drain excess carbon from the atmosphere. Building another thousand prayer mills, or a thousand thousand, will not spare the world the hardships slouching toward us. Not even a billion of them will alter the pH of the oceans or reverse the shrinkage of wild habitats. But the builders of the prayer mills, if they believe in their actions ardently enough, may feel in their lives a level of peace that the rest of us must view with respect, if not envy. In the end, what we know for certain is that Sapiens are good at fiction. They often believe in bunk, and sometimes the bunk works for them. At which point, the bunk becomes something that is not bunk. Perhaps it becomes a species of truth.

All this is neat and tidy—it offers an accommodating way to think about faith healing, and other forms of magic, which are otherwise not easy to defend. But it is not complete. Not all bunk is as innocent as the prayer mills of Shey. No matter how many endorphins a snort of powdered rhinoceros horn may release in the brain of an aspiring Romeo, the murder behind that fantasy remains repugnant and wrong. The same goes for tiger bone and bear gall, whether obtained from "farms" or from the wild, and for the harm done to countless other creatures in the name of Sapiens' health and happiness. In this regard, the world of bunk mirrors that of science, which has provided humankind not just the means to understand the world but the leverage to place it in a state of perpetual crisis. It may be stating the obvious to say that neither science nor bunk is exempt from moral stain, but there you have it.

43

Reverie

I WONDER if our knowledge of natural selection and plate tectonics leaves us more connected to Earth or less so. The material influences of modernity have clearly diminished our connection, but the effect of a scientifically informed worldview is harder to discern. Perhaps our growing omniscience makes no ultimate difference. Even absent knowledge of how our world came to be, our immoderate success as organisms on Petri Earth might still have pushed our biological hegemony to the farthest limits.

Certainly the tools we call "technology" lengthen the levers with which we manipulate the world, but who can say if the lessons of the Galapagos or of the seafloors strengthen our use of them? Perhaps the great Earth theories constitute a different class of augmentation. If they at least pry loose the gift of beauty and amplify the music of the spheres, then we may rejoice to have them.

Meister Eckhart, a Christian mystic of the late thirteenth and early fourteenth centuries, never claimed that God spoke to him. Through a lifetime of contemplation, the most he admitted was that, once or twice, he may have heard God clear his throat. Many admirers of science have said they sense a similar *ahem* in the structure of the atom, the composition of a distant nebula, or the iridescence of a hummingbird.

Sometimes science only elaborates what indigenous people have always understood. The Yurok of the Klamath River, for instance, hold that the redwood forest is no mere sum of trees, plus animals, plus understory plants, plus soils. It is a single living wholeness that embraces the entirety of the forest from the giant trees down to the smallest pill bug. Scientists came late to that understanding, which now lies at the heart of community ecology. Still, knowing something of how the redwood community actually functions—understanding, for instance, the water flux of the great trees, how the giants literally drink from the fog that rolls in from the Pacific, else they could not grow so tall—only adds to one's awe. Beyond every dimension that we see, lies another that we do not. With the naked eye we see the starry night sky, but with a telescope, we see light beyond light beyond light.

There is beauty in understanding our true location in the cosmos, as small as we are and as big as we sometimes feel. Grandeur abides in this view, the grandeur of a universe suffused with mystery, of which we perhaps understand enough to appreciate, more fully than ever our ancestors could, the majesty of the whole.

44

The Heist

IN THE gompa, the lama's breathless delivery causes Wangmo and Prem, our interpreters, to struggle to keep up with his account of the robbery. There is so much to tell. And it happened so fast. The thieves came in the night. They dragged the lama from his house beside the gompa and beat him with rifle butts, breaking two of his ribs. He says he did not tell them where the relics were hidden; they seemed already to know, which, if true, means that the theft was an inside job. You could count on two hands the number of people who knew about the secret crypt. But perhaps the lama was protecting his pride. Perhaps, when they would not stop hitting him, he gave away the secret. Truth can be lost in translation, or in other ways.

This was in 2015, and it was not the first robbery of the gompa. Four years earlier, another break-in prompted the lamas of Shey to gather the precious statues and figurines that remained, some of them ancient and adorned with precious stones, and to hide them in the farthest corner of a storeroom. Then they built a wall of stone and mud, the same as any wall in the gompa, windowless and doorless, to seal up the treasure. A stranger who went into the storeroom would see a dim, blank wall with all manner of junk leaning against it and think that it marked the limit of the building, that there could be nothing behind it. That was the intention.

These days, more and more outsiders come to Shey, not just the devout but also fortune seekers. Many of these outsiders come to harvest *yarza gunbu*, a weird fungus that invades the head of a particular caterpillar soon after it hatches in the tundra grasslands. The fungus consumes the unlucky caterpillar and erupts upward through the thin soil, to produce a miniature tower, only a centimeter or two high, that with a certain amount of imagination can be seen to resemble an erect penis. The yarza hunters pour by the thousands into the high country as the snow recedes in the spring. Crawling on hands and knees or shuffling stooped across the damp heights, they stare intently at the ground, straining to spot the flagpole structure of their quarry. Gathered and dried, these rather unappetizing avatars of the male principle sell at cocaine prices as a remedy for impotence and as a general tonic for health. Their market includes a large swath of Asia, especially China. Some call it "Himalayan Viagra."

Outsiders coming to Shey for the yarza gunbu harvest would have known that the gompa contained treasures; they would have learned the trails of the region; perhaps they might have made useful local contacts, and they might not have been as loyal as locals or pilgrims to the dicta and traditions of the lamas.

It was late April or May, the height of yarza gunbu season, when the thieves attacked. There were at least three of them. Besides beating the monk, they tunneled through the concealing wall and stole twelve statues, five big ones and seven small, the best of which were inlaid with quantities of coral, turquoise, and other gems that worshippers had offered as gifts over the centuries. The thieves packed their loot on horses and rode away eastward. Some say that at the next trail junction they turned south for Dho Tarap.

Amchis, traditional Tibetan healers, came from Bhijer and Saldang to minister to the wounded monk. He reports he is well now. He says he has received many blessings.

He also expresses gratitude for the funds raised by Roshi Halifax soon after the robbery. The money enabled the purchase of an immensely heavy Chinese safe in which the lamas might keep such treasures as the gompa still possessed. A helicopter delivered it to Shey last summer, but, sad to say, it could not deposit the safe close to the gompa, there being no area large and flat enough for the helicopter to land. The chopper had to unload its ponderous cargo some distance away, for which reason the monks were obliged to hire fifteen porters, at a cost of three thousand rupees (about thirty dollars) to carry the safe to the gompa. The unexpected expense was a very great burden, like the safe itself, but now it is installed, and the treasures of the gompa are within it. The lama further asserts that soon the stolen statues will also be placed in the safe. It is certain, he explains, that the statues want to come home, and they will make the thieves sick until the miscreants bring them back.

The monk delivers this prediction while standing in the cold gloom of the storeroom before the broken wall of the crypt. The hole whence the statues were extracted gapes at us. There is a silence. Someone, changing the subject, asks if George Schaller and other Westerners have paid a recent visit. They should have been in Shey about ten days earlier. Wangmo translates the question. The monk breaks out in laughter. He speaks raucously, and gestures left and right. Yes, he says, he knows Schaller! And now Prem and Wangmo are laughing too. Wangmo says, "He say, that man is a big headache. Always wanting to know, does snow leopard sleep here? Does snow leopard sleep there?" The monk is now doing stand-up comedy. He contorts his face in an expression of satiric incredulity. Such questions! So many of them! So annoying, so absurd! Everyone has a good laugh at Schaller's expense, grateful after the sad tale of the robbery to enjoy this fresh evidence of Westerners' lunacy.

Eager for the sun, we follow the monk to the roof of the gompa, climbing a succession of thick timbers notched with footholds—a Dolpo stairway. We emerge into bright sunlight. Overhead, the ever-present choughs dip and glide on a steady wind. We ask more questions about wildlife (but fewer than Schaller would have asked), and the monk answers tolerantly but also with growing impatience. Our interests seem trivial. Snow leopards, bharal, wolves—they are no more noteworthy than the weather. Every day it is clear and cold. The animals come and go. That is the full story.

But then the word "yeti" is uttered, and the monk's energy surges anew. He rattles on for some minutes. Wangmo cannot keep up. Only scraps come through: male yetis don't come around here very much. They seem to like it better up by Mount Kailash. In this area one tends to encounter only female yetis.

Why is that?

Because it is hard for the little yetis, the young ones, to cross the rivers between here and Kailash. And so the mothers stay behind with them.

Do you see them often?

Not so much when the blessings flow. But when things are amiss, when disharmony afflicts the spirit world, yetis tend to prowl. And there can be no mistaking that the present time is not good. Witness the robbery of the relics, the violation of the gompa, the statues yearning to come home.

We gaze out from the chest-high parapet of the roof. The valley of Shey and its cobble plain spread below us, shimmering in brilliant light. We face southward, toward Kang La. The stream that drains the pass, which Matthiessen called Black River, sparkles toward us. A portion of its water enters the narrow channels leading to the prayer mills. Inside the stone cabinets, the water pushes the fins that turn the axles, and the cylinders inscribed with mantras spin their hopeful prayers into the cosmos. *Om mani padme hum.*

45

Hope

HOPE MEANS different things to different people. In its simplest form, it expresses a desire for things to turn out well, for a worrying story to have a happy ending. Most of the time when people ask about hope, they are asking, will everything be all right? Can we return to how things used to be, when this worry did not exist?

With regard to climate change, the answer is *no*: too much CO_2 and other heat-trapping gases already burden the planet's atmosphere and oceans; the effects will be with us for at least several lifetimes; we cannot draw a get-out-of-jail-free card, for none exists.

But hope has other meanings. Václav Havel wrote, "Hope is definitely not the same thing as optimism. It is not the conviction that something will turn out well, but the certainty that something makes sense, regardless of how it turns out." Havel was linking hope to the philosophical distinction between instrumental good and intrinsic good. Something is instrumentally good if it produces a desired result. Its goodness depends on outcome. But a thing is intrinsically good if doing it is virtuous of and by itself—that is, if its value exists independent of result. The essence of hope, Havel was saying, is to believe in the intrinsic goodness of right action. Through his many years as an outcast or in prison, fighting the Soviet domination of Czechoslovakia, Havel never knew (until the Soviet Union's final collapse) if his efforts would succeed. Yet he persisted in a spirit of hope, knowing his course was correct.

The novelist Barbara Kingsolver also distinguishes between hope and optimism. In her view, "The pessimist would say, 'It's going to be a terrible winter; we're all going to die.' The optimist would say, 'Oh, it'll be all right; I don't think it'll be that bad.' The hopeful person would say, 'Maybe someone will still be alive in February, so I'm going to put some potatoes in the root cellar just in case.'" Kingsolver concludes, "Hope is a mode of survival: I think hope is a mode of resistance." The hope she describes is close to the ecological notion of surprise: that sometimes big, consequential things happen with virtually no warning—an earth-quake or the fall of the Soviet Union being good examples. To trust in the uncertainty of the future, believing in the possibility, however remote, of beneficial change—this is the essence of hope.

Of course, surprise is no panacea: it can harm as well as benefit, a new coronavirus triggering a pandemic being a salient case in point. Surprise comes to us out of the vastness of what we don't know. It is amoral and uncaring. But it is also central to true hopefulness. Roshi Joan puts surprise and uncertainty at the center of her teaching. Placing trust in "not-knowing," she says, offers a strategy for dealing with dark times: change is certain and there is always a chance things will improve. Here is where Kingsolver's wisdom connects with Havel's. Kingsolver is talking about future surprise, the uncertainty of how the winter that lies ahead will turn out; Havel is talking about how we carry ourselves in the meantime: we have to "do what makes sense," irrespective of outcome. In jail Havel could not know if the Soviet Union would crumble during his lifetime. Nor could Nelson Mandela, during thirty-three years of imprisonment, know when apartheid might similarly disintegrate. But when the long-desired surprise arrived, both men, having done "what makes sense," seized the moment and helped render the surprise as beneficial as possible. The essence of their prepa-ration was that they never lost hope.

46

Tsakang

IN *THE Snow Leopard* Peter Matthiessen arrives at Shey Gompa, the object of his pilgrimage, only to find the door of the hermitage locked. Worse, the lama presiding there, whom he has been "so anxious to find," is absent, and no one will tell him where he has gone or when he might return. The lama is *tulku*, a reincarnated one, the flesh-and-blood materialization of a spiritual entity that has taken a succession of human forms over many generations.

Days later, Matthiessen follows a trail westward from Shey to a hermitage called Tsakang, the "Red Place," where several spare stone buildings hang on the face of "bright cliffs of blue and red." The only flat area Tsakang affords is a sun-drenched ledge where Matthiessen encounters "two bronze-skinned monks." The younger of the two mends a pair of woolen boots, and the other, "curiously ageless, is a handsome cripple in strange rags of leather." They exchange amicable but scarcely audible greetings, while the older monk spreads "a yellow mix of goat brains and rancid yak butter" on a goat hide. Soon Matthiessen takes his leave, and "the two figures bow slightly, smile again, and keep their silence."

Midway through his sojourn at Shey, Matthiessen learns that the crippled man who was curing the goatskin is the lama of Shey. This revelation provides the thematic center of the book. Matthiessen visits Tsakang at least twice more. He learns that the lama once freely roamed Shey's trails and mountain slopes, but some degenerative process, possibly arthritis, deprived him of the use of his legs. Unable to return to Shey, he has been confined for eight years to Tsakang. In all probability, he will die there.

Shortly before his departure from Shey, Matthiessen makes a final visit to Tsakang. Jang-bu, the porter who comes with him to serve as interpreter, "seems uncomfortable with the Lama or with himself or perhaps with us." Because of this, Matthiessen tells him not to ask questions about the lama's isolation and debility. Yet Jang-bu does so anyway. Jang-bu boldly asks the tulku if, crippled as he is, he can be happy.

And this holy man of great directness and simplicity, big white teeth shining, laughs out loud in an infectious way at Jang-bu's question. Indicating his twisted legs without a trace of self-pity or bitterness, as if they belonged to all of us, he casts his arms wide to the sky and the snow mountains, the high sun and dancing sheep, and cries, "Of course I am happy here! It's wonderful! *Especially* when I have no choice!"

Through his long journey across Dolpo, Matthiessen has longed to see the snow leopard, has felt the presence of the elusive cat, and has spent many cold, stiff hours at lookouts hoping one will reveal itself. But to no avail. He has seen "blue sheep dancing on the snow," but he fails to glimpse the leopard of the Himalaya, and he never will. His visit to Tsakang helps him find no fault with this. He ends the account of his visit to the crippled lama:

Have you seen the snow leopard?
No! Isn't that wonderful?

47

Sacred Rage

THE TRAIL from Shey to Tsakang carves a furrow in the side of Crystal Mountain, one of many scars gouged along the contour of the slope. Dozens of ruts run in parallel through the sorrel and cinquefoil, carved by the hooves of yaks that have plied the route for centuries, not to mention the humans who guided them and the bands of horses, sheep, and goats that also straggled along. A white flag flutters from a chorten where the main trail tops a ridge. The chorten, like a buoy at the mouth of a harbor, marks a channel in and out of Shey. It tolls a silent message, chiding all who pass by to strive toward the Buddha mind. The painted eyes at its base perpetually interrogate, Are you striving?

Beyond this chorten stands another, and another beyond that, buoys of a channel both physical and spiritual. Upslope on the mountain and yonder across the canyon, flags flitter from outcrops and knolls, each scrap of cloth a continuous windblown prayer. The land trembles with memes: be mindful, be present, be compassionate, be not dismayed.

I feel examined as I pass the chorten. I feel I flunk. I will always flunk. I envy those, like Matthiessen and Roshi, who find what they seek in Buddhism, but I can never be a Buddhist, still less anything else. I tell myself that I object to key points of religious doctrine. Tibetan Buddhism's concepts of karma and reincarnation, for instance, offer a fairy-tale alternative to the randomness of the universe. We want to believe that every effect has a cause, that virtue and good works will be rewarded, and that fortune in life is not a casino game. As Sapiens we deftly conjure stories to soothe our doubts, and it is no accident that the best fictions are also useful for social control: mind your betters, or you'll come back as a dung beetle.

But Zen Buddhism, Roshi's discipline, carries none of that baggage. So I cannot say it is doctrine that keeps me at arm's length. Nor is it the hierarchical structure of Zen Buddhism and the appurtenances that go with it—the costumes, ritual, and priestly jargon. It is not even the pressure to conform, so much as the fact of conformity. Roshi acknowledges that "every religion is a kind of club," and I think my resistance lies there. I am not a joiner of clubs. Ages ago, a high school teacher called me aside to tell me that I was a rebel without a cause. To this day, I do not know if he intended praise or criticism, nor even why he thought the message important to impart. Perhaps he merely meant to state a fact. In any event, it is not a fact I would contest.

And so I feel at odds with the chortens by the trail, as I trudge toward Tsakang on a bright day, wearing my atheism like a coat of quills. Rounding a turn and glimpsing the red hermitage clinging to its rocky cliff, I wonder if I have raised my defenses because I am approaching so holy a place. Centuries of monks, including our amiable companion Lama Ngawong, have so filled Tsakang with their chants and prayers that, even from a distance, their devotions seem to coat the hermitage and its canyon with a sacred dew. In this place they have shed old selves and emerged renewed. Arriving at the red walls, I set down my pack on the ledge where Matthiessen met the crippled lama: *Of course I am happy . . .* Especially *when I have no choice!*

Unhappily, I go in. A massive staircase fills the first room of the main hermitage. Its woods are tight-grained, thick, and burnished, its design a giant puzzle. The treads and risers are mortised into the side beams, so that all the pieces must have been joined in one simultaneous and perfect fitting. Slats pierce the beams to lock the mass together, and wooden pins secure the slats. Every joint is precise and flush. The whole structure, without the least intervention of metal, stands unbending and mute, not creaking under the heaviest tread. This work of carpentry leaves no doubt of the devotion that inspired it. It is a prayer of wood, of months of artful, meditative diligence.

A puja has begun in the shrine room at the head of the stairs. The resident monks, Roshi, and half the members of our expedition have crammed inside, hip to hip, until not an inch of open floor remains. The passage to the door is jammed with people craning forward. Chants fill the building.

I climb a ladder through a hole in the roof, and come out where the sun feels warm and the air tastes like ice. The hermitage vibrates from the chanting below, and the drone of voices drifts into a swallowing wind.

The rooftop abounds with the paraphernalia of faith. Strings of prayer flags flap along the parapets, and from a mast in one corner, a multicolored pennant trails streamers of red and blue. Other spindly masts are capped with brass finials and adorned with pleated yellow skirts. And still another flies a cloth cylinder checkered in blue and white that shimmies in the wind. At the center of the roof a tall, narrow table, crudely made, supports a tower of eight ceramic rings, each different, of graded sizes, evoking a skeletal mini-chorten, another representation of the mind of the Buddha, another meme presented to the cosmos.

And the cosmos looms. Beyond the crumbling stone parapet, and far past the reach of the susurrations rising from below, the ocean of the Himalaya rears up on every side, its peaks knife-ridged and snow-streaked, shining beneath fleets of cotton clouds.

Squinting, I can just make out the gompa at Shey, stolid on its lonely bluff, a dark pinprick on a tawny mountain. Perhaps the intoxication of the wind or the wild spaciousness of the mountains has got into me, but the crippled lama who had no choice seems fleetingly present. His ledge is just below; I look over the parapet and see it now, the flat stones and wisps of grass, the place where extremities of spiritual strength and physical weakness reconciled.

Strive to cultivate happiness, I say to myself. Especially when you do have a choice. But, for me, in this moment, the idea is just words. The mountains loom, more real than thought. So quell your thoughts, I say. Be still in heart and mind. Let all things pass through. And now I feel the wind anew, the sun's radiant warmth. I hear the intoned mantras welling up through the roof hole. I focus on my breathing.

Soon I look down again at the ledge of the lama of no choice, and imagine his approval of a conversation for which a friend of mine was present.

The Dalai Lama was having lunch in the cafeteria of an American ski resort. His Holiness had come up the mountain amid an entourage of monks and Western acolytes to enjoy the snow and sweeping vistas. The wintry high country reminded him of Tibet. You may suspect that the spirit of the group was muted; the Westerners, in particular, were overawed. They ate in near silence. But not so a young woman who worked in the cafeteria, bussing tables. She had tangled yellow hair and a bold face. She came over with her tray. After a moment surveying the group, she asked, "Um, are you the Dalai Lama?"

"Yes," he smiled. "I am."

"Mind if I ask you a question?"

"Please."

"What is the meaning of life?"

Everyone froze. It was the question no one had been brave enough to ask.

The Dalai Lama beamed his famous, impish grin. "Oh, that is an easy question," he said, and everyone held their breath for the answer. "The meaning of life is to be happy," he said. He paused. The tone of his voice became almost confidential as he inclined toward the young woman. "Now, how to be happy? That is the hard part."

She nodded, "Thanks," and began removing the empty bowls and plates. From around the table came the sighs of breath released.

The lama of no choice partook of this wisdom, which no one learns easily, not at its deepest level, least of all me. I look out at the mountains of Shey. Hard-used for centuries and worse-used since the old balances of netsang succumbed to a closed border, the slopes have been grazed to a fare-thee-well. After the yaks and sheep and goats take their considerable share, there is little left for the bharal, and even less since yarza gunbu has brought more people and more livestock into higher country. And eventually, perhaps, there will be too few bharal to support the snow leopard. Even the dome of the sky, sick with carbon, has lost its impervious magnificence. It is grand and beautiful, but also strangely and newly vulnerable. The same may be said of the unseen glaciers of the high peaks, which reflect the sky's dolor as they shrink and withdraw. My heart is suddenly at odds with the spirit of Tsakang, for anger is rising within me. Even here, in this lonely, holy place sheltered from the turmoil of the world, the fetor of Petri Earth taints the air.

And then I think, doubling down on my apostasy, if I were the lama of anything, I would cultivate rage.

Not rage in the usual sense of uncontrolled, diffuse anger, but an opposite kind of rage, focused and deliberate. It would be fueled by calamity and by the now increasing and inescapable unhappiness born from the end of ice, from rising and overcooked seas, and from jacked-up tempests and climate instability. It would be a rage amplified by knowledge that the worst of these woes were begot by deceit and willful ignorance. Roshi's adjuration notwithstanding, there is much fault to be found in the present. All the necessary facts to address our planetary plight have been known for thirty years and more. The leaders of nations and powerful companies knew them, understood them if they chose to, and not only ignored them but actively campaigned to suppress and invalidate them. They were abetted in this by a sufficiency of lackeys and politicians—the best that money could buy—who forestalled adoption of the corrective policies and technologies that lay entirely within reach. Their behavior was not less immoral than that of cigarette companies marketing cancer or of opiate manufacturers selling yet another brand of addiction. Their cynicism has been hardly less immoral than flying jetliners into office towers. Murder takes many forms and may be variously delivered, but if the act is willful and informed, its moral valence does not change.

Part of my rage is that I am complicit in this crime. I drive a car. I fly on airplanes. The solar panels on my roof do not erase the thousand things I do and buy that pump waste gases into the sky. All my kin, my closest friends, my Nomad companions, the woman I love—each of us is complicit, entangled in a web from which we would escape, a web from which the route out has long been known and achievable save for the purchased politicos, the disciples of profit-at-all-costs, and the inertia and gullibility that infect us all and that are native to our character as Sapiens.

I cannot deny this rage. I have studied the "Thirty-Seven Practices of a Bodhisattva," composed six hundred years ago by Togme Sangpo, a Tibetan monk. It is widely taught as a guide for spiritual development. Near the top of the list, at number two among the practices, is the imperative to "leave your homeland." The intent is to minimize attachment to people and place, the better to transcend the unnecessary longings of the material world. But such an injunction, if followed comprehensively, would leave people and places undefended, a surrender inviting plunder and abuse. The rage of which I speak is an opposite response. It arises from loyalty to place. It is a rage that pulls one, not into a hermitage, but back into the lists, a rage that is cool, like the spirit of the monk, a rage that the western American writer Terry Tempest Williams calls sacred: *Sacred Rage*. The proper discipline would be to channel the rage, not simply vent it like steam from a relief valve. Instead it must be focused and made to do "work," like the steam that drives a turbine.

Big words. I laugh at my flight of mental bravado. I imagine the stoical mountains laughing at me too. Ha-ha! If nothing else, I can serve the world as a walking, breathing, animate joke. Am I up to the task of marshaling my rage? Probably only in a mediocre sort of way, but probability justifies nothing. Were I to follow my rage and put it to work, might I do so in a state of happiness, unattached to distant outcomes? Might I focus it, like a good hospice worker, on the intrinsic good of making each day, each hour, useful and positive? The odds are at best only fair, but perhaps that is what the chortens have been asking me to do. Perhaps they are not meant to say the same thing to every passerby. Perhaps they lead different vessels to different harbors. Perhaps I should attend to them in a new way.

48

Kang La

THE ROTUND lama beside whom I am riding has twice my girth, and I am not a wraith. He rides heavy and immobile, and, alas for his puny mount, which is smaller even than mine, he makes no effort to shift his weight in rhythm to his horse but sits his saddle as he would a fallen log. A woman toiling up the trail on foot, perhaps his wife, huffs beside him, a large wicker basket laden with camp gear strapped to her back. A second attendant, a young man similarly burdened, has gone ahead to build a fire and prepare midmorning tea. Yesterday the lama entertained us jocosely in the cave where the monks of Tsakang draw their water. Now he chants as he rides, half singing, half mumbling, "om mani padme hum" in ceaseless repetition, pausing only to yank his horse's reins, apparently without purpose. The reins are entwined in his fists along with a string of prayer beads. He chants and yanks, chants and yanks, like a drummer who keeps two beats. He yanks even when his animal wants only to pause and have a proper shit. The monk's cassock is sashed at the waist and dyed the same reddish brown as the plaster of the hermitage of Tsakang, where he is keeper of the key. On his shaven head rides an ageless wool cap that would not look out of place on a friar in Chaucer's time, trudging toward Canterbury. The monk, like the rest of us, is bound for the high pass of Kang La, and beyond that, for Phoksundo and eventually Kathmandu. His chanting earns him merit toward the next life, while his miserable horse plods on.

I rode horseback on the approach to two other passes, the better to conserve energy for the final climb to the summit and, in theory, clarity of mind for taking notes. Pau Lama leads my horse, as actual riding is discouraged. Pau (pronounced *Pö*) is twenty-four, married, not yet a father, and meticulously polite. He is also indefatigable. Even on the steepest climbs, he sings as he walks. Like many others among the crew, he hails from Humla, home to his kinsmen Prem and Tenzin. Pau speaks Tibetan, Nepali, and Hindi, as well as scraps of English and Mandarin, one or both of which he hopes to brush up on during the coming winter. He spent the summer escorting Indian pilgrims, group by group, in koras around sacred Mount Kailash. Such trips are the bread and butter of Tenzin's company, Sunny Treks.

Most koras of Mount Kailash are accomplished in two days, the length of the standard trip that Pau guides. Clients arrive by plane or helicopter, virtually from sea level, and commence their trek the following morning. Along the way they summit a pass higher than Kang La, rarely pausing to acclimatize. Altitude sickness is common. So is garbage. Each year ten thousand pilgrims ply the formerly pristine route, where plastic bottles and empty noodle packets now abound. Pau says he much prefers an expedition like ours, and I do not doubt that this is true, but it is faint praise.

Our route takes us south, up the stream that Matthiessen called Black River, the proper name of which is Kangju: Snow Waters, as Kang La is Snow Pass. Crystal Mountain is on our right, and the morning sun now bathes its high, bare slopes in brilliance. We cannot see the summit. It occurs to me that in our several days at Shey I never saw the top of Crystal Mountain, and this saddens me. Perhaps I glimpsed it when we approached Shey from Namgung, before we descended to the valley floor, but I would not have known what I was seeing. One morning I climbed Somdo Mountain, which rises behind the gompa, but only far enough to reach the site where a clutch of vultures had fed on a dead marmot. Even then, I could not see the top of Crystal, and I did not climb farther. Instead I returned to camp and rested, later wandering up the braided channels of Kangju to watch white-winged redstarts flit through the willow scrub.

Every evening at Shey, I marveled at the arrival of three or four long caravans of yaks bearing beams and boards from the forests of Phoksundo. The whistles and shouts of the herders, the hollow-toned tolling of bells on the lead yaks, and the grunts of the bovines as they shuffled down the trail made a music that still haunts my memory. I wondered how long the forests of Phoksundo might hold out before the last mature trees are felled and hewn. And I wondered, too, whether it mattered. A warming climate likely dooms those forests, as it dooms the pine forests of home, to drought, insects, and fire. Nothing lasts, and where humans and their emissions are concerned, impermanence accelerates.

One evening several caravans arrived snow-dusted from the slopes of Kang La. The last to pull in was guided by a man I had met in Saldang, where we had talked of snow leopards and wolves and the loss of livestock to predators. In three days since then he had traveled to Phoksundo and beyond, loaded his animals with lumber, and returned more than half the distance homeward. He must have spent every waking minute in motion. Beside him I felt a slacker. His yaks now stood immobile, waiting to be freed of their burdens. He approached the nearest animal, seized a rope, and with one strong pull released a single knot, so that the whole load of beams and boards clattered to the frozen ground. Then another knot released the packsaddle, which he tossed to the ground. The yak, grunting, shambled toward the river. He did the same for each yak, and slowly built a small mound of packsaddles. When the last animal was unloaded, he spread a tarp over the saddles and weighted its edges with cobbles. Tomorrow, he says, he will water his beasts in Saldang, and perhaps the next day, he will take them past Shimen to the door of China. I marveled that his eyes betrayed no hint of weariness or strain.

Now we climb the trail the Saldang trader descended two evenings ago. The icy waters of Kangju murmur beside us, as we travel in deep canyon shade, and the air bears the crispness of winter. Light snow fell in Shey the last two nights, and before that it fell on the sorrel ridge where we camped above Namgung. We can be sure that the snow pass of Kang La will live up to its name.

If Buddhists hold Crystal Mountain sacred, it stands to reason that followers of Bön might revere a nearby mountain too. Pau confirms that this is so. On our left, opposite Crystal Mountain, rises a singular mass of tortured rock. It is a naked knob two thousand feet tall, straight as a skyscraper, and bearded with icefalls. The wind has polished its metamorphic swirls to a high, dark gleam. Perhaps when touched by afternoon light, its black face will lighten, but for as long as the morning sun is low, one needs no leap of imagination to picture a gang of demons lurking in its recesses.

The trail rises steeply after the Bön megalith, lifting us at last into the day's light. We cross the lip of a broad bench, beyond which smooth low ridges recede toward a natural amphitheater, a cirque, cupped into the headwall of the mountain. This immense hollowed space was once the seat of a glacier. Past the last wavelike ridge, a glistening white remnant of the glacier comes in view, scabbed to the base of the headwall. Below it, at the bottom of the cirque, glares the dark eye of a shallow tarn, which Matthiessen called Black Pond. Today is the thirteenth of October. In 1973 Matthiessen and Schaller, traveling the opposite way, descended from Kang La and arrived here on the thirty-first. They encountered much snow. Their porters dropped their packs and pushed ahead in darkness, seeking the stony comforts of Shey, but Matthiessen and Schaller camped beside the pond and made a meager dinner of their last tin of sardines. They passed a desolate, sub-zero night: "GS is a remorseless sleeper," wrote Matthiessen, "but for me the night will be a long one. . . . though I wear everything I have, I quake with cold."

We continue upward, past Black Pond. Patches of snow streak the walls of the cirque like careless dabs of paint. The trail zigzags through tightening switchbacks. I am walking now. Pau has given the horse to another Sherpa to lead and has dropped back to assist a hiker who appears to be struggling. I steal a glimpse of the horses and people behind me and see a woman in a red jacket—I cannot tell who it is, for she, like everyone, is bundled against the cold. She clutches the back of Pau's jacket with a tight fist and follows him as a blind person might, head downcast. Perhaps the heights and large spaces upset her, for the sweep of the land now staggers the mind. With each step upward, the boundaries of the snowclad world recede, and we arrive in the realm of the sky. A bitter wind roars. I am swathed in three layers of pants, five layers of shirts, plus parka, neck gaiter, two pairs of gloves, and a hat cinched down by a strap beneath my chin. The wind snatches at the fabrics, flapping them noisily at a hundred points. I am an irritant to the air, an obstruction this mountain would quickly shear away. We are seventeen thousand feet above the sea, and a lungful of air now delivers half as much oxygen as at tideline. The pass towers above us and we inch our way up a deep, wide snowfield, the trail a trough between white walls. I walk in a stoop, bent over the hiking poles that have become my forelegs. I am an insect scaling the wall of a jar. Immediately in front of me is a horse, led by Jigme, I think, but I cannot see past its bulk. The trail steepens, tilting up and up until the horse is more above me than in front. I am glad for a reason to stop and enlarge the space between us. With every gulp, I feel I bite the air. It enters me like food, and I am ravenous for it. I begin again: Step with the left foot. Inhale. Plant the right pole. Exhale. Plant the left pole. Inhale. Step with the right foot. Exhale.

I kick steps in the snow to gain purchase. The footholds stretch higher and farther apart. I am a mouse on a staircase, every tread too high. Above me the horse, panicking, lunges upward. It misses its

landing and staggers back, rearing on hind feet. Its legs tremble vio-
lently. It teeters within a degree of falling backward. If it falls, it will
either crush me or plunge from the trail and skid down the snowpack to
shred itself in the rocks. Perhaps I will go with it. But Jigme pulls hard
on the rope. The horse cants forward. Its forehooves find the trail. It
snorts and lunges again, clambering to a ledge. We are almost at the top.

Inhale. Plant the right pole. Exhale. Step. Inhale. Plant the left pole.
Step. All I see now is a wall of snow in front of me. All I can do is move
my hands one at a time, my feet one at a time. And breathe. My mind
is wonderfully clear, my body at its limits. Lungs heave. Legs are leaden,
close to collapse. Inhale. Exhale. Plant the pole. Step. I suddenly sense
my innards loosening, my organs becoming liquid. I wonder wryly if I
can control them. Only one thing exists in this world. It is this trail, this
snow, this air, this disobedient body, all bound together. The moment
is a unity. Breathe, plant, step. Breathe, plant, step.

The last of the trail punctures an ice cornice. Snow at my shoulders.
I kick steps, climbing the slot, as though through a manhole. Then
my head is above the snow. And the wind blows the world back to far
horizons: I see snow-swept peaks, gray-bottomed clouds, space that
yawns to infinity. One more step, and now I stand on the slender ridge
that is the top of the pass: Mountains! Wind! Clouds! *Kiki soso lha gha
lo!* Friends are there, all shouting. I turn in a slow circle, and in every
direction I see the majesty of Tibet and the high Himalaya. I feel as
though my path has been leading to this place forever. All around me,
brilliant in the light of the sun, I see the world resplendent.

49

Birches

FOR THE past month we have been living in a moonscape, where the tallest vegetation could not hide a golf ball, where the furnishings of the land consisted of glacial outwash and the rubble of fallen cliffs. Perhaps Dolpo's nakedness accounts for its spirituality. Nothing is screened from view. What is there, is there. The world, says Dolpo, is as stark as sunlit boulders, as clear as snowmelt. Make your mind the same.

The snowless, south-facing descent from Kang La sustains the lunar feeling. It began through a mile of fine gravel, grains of stone the size of lentils, just large enough that the wind could not carry them away. But the wind had tumbled and rounded them so that they rolled like ball bearings. The gravel slope leans at a giddy tilt, the angle of repose, a vast dune. You take a step, and the footing gives way so that you slide down another foot. Add a light-footed hop and the effect is like skiing, a dance with gravity. The stony world has become soft, the labor of the climb is now replaced by near weightlessness, the mountain inviting us to bound along its side like astronauts. We are children again as we laugh our way down a thousand feet.

The gravel scree ends at the head of a chasm where darkness pools. We enter the defile, stairstepping on boulders, descending as though down a drain. We leave the bitter wind above us and stop to lunch in a widening of the gorge. Our pack stock scatter to graze. Then onward we press, crossing and recrossing a rivulet that froths down the rocks, gradually gathering other rivulets from either side until it has swollen into a torrent.

Our descent roughly equals the walk from rim to river in the Grand Canyon, but it is steeper, for while the amount of altitude lost is the same, our trail is a mile shorter than the canyon's most direct route. The scenery is comparable too, in beauty and ruggedness. But where the Grand Canyon impresses with the age of Earth, the gorge below Kang La impresses with its youth. Everywhere lies evidence of rockfall and landslide, cliffs gleaming where they have fractured, their rocks newly exposed and unweathered. We are in a terrain still being born, a neonate geography.

Guidebooks explain that the Himalaya formed 55–40 million years ago (mya) when India, having barged into the Asian plate 10 million years earlier, renewed and intensified its push, forcing a mountain range to crumple upward along the line of contact. Those first mountains were low and rounded. The terrestrial biota of Asia, even including its frogs, had no difficulty roving over the modest new heights and into the beckoning habitats of India. As birds evolved, they would have glided over those early mountains, ancestral cranes (which appeared 56–34 mya) included. Today's Himalaya, the modern barrier to animals, armies, and even the atmosphere, did not arise until recent geological days, in the late Miocene (11–5 mya) when the mountains rose tall enough to stall the movement of weather systems. Only then did the south Asian monsoon assume its role as arbiter of life on the subcontinent.

Even then, the Himalaya was not the august chain we see today. The last 3–4,000 meters of its height (more than 10,000 feet) were added in a geological yesterday. It began in the early to mid-Pleistocene, about 1.6 mya. The rate of upthrust started fast and then grew faster, peaking about 800,000 years ago. The mountains are still at it. The Nepal Himadri—the highest of the Himalayan ranges, which runs the length of Nepal and includes Everest, Annapurna, Dhaulagiri, and other fabled peaks—appears to be growing at an annual rate of 7 millimeters (+/- 2), which amounts to an inch every three or four years. Dolpo lies north of the Himadri, wedged between it and the Tibetan Plateau, in a zone called the Tethys Himalaya, named for the Tethys Sea that formerly separated India from Asia. We have been treading the uplifted sediments of that sea virtually every step of our journey. If anyone has calculated the rate at which this lonely portion of the Himalaya is rising, it is not widely known. But the rate is not zero.

Such facts, by themselves, are surprising. Even more startling is that we know them. Based on the age, volume, and mineral content of sediments washed out from the Himalaya into the Bay of Bengal, researchers have reconstructed rates of erosion for the mountain chain. Because erosion is proportional to the rate of uplift, they have deduced the history of mountain building. We live in an age of marvels, not just of toys, tools, and gadgets, but of insight. We know that the Himalaya is young and dangerous. Earthquakes arising from the mountains' tectonic restlessness have killed nearly 150,000 people since 1980, most of them in the terrible Kashmir quake of 2005, with another 9,000 victims in Nepal in 2015. We also know that more such disasters are inevitable.

Late in the day, the defile becomes less steep, and the trail eases. I begin to see small gravels and then sand at the margins of the stream, deposits that the torrent has failed to wash away. I round a massive pile of boulders at a bend of the gorge. Ahead stands a steep, water-cut bluff on which lie patches of something dark amid the glint of bare stone. Pockets of soil, I realize. And something else, spindly and angular. I look harder: vertical sticks. From one pocket of soil, they branch up to a height of four or five feet. The sight takes a moment to register: it is a tree!

Aside from the forlorn willow in the Saldang schoolyard, I have not seen a tree in a month. The moment seems a homecoming, and I feel a wave of delight. Beyond the first tree are others, spreading back along the soil-splotched bluff. They are birches, with peeling, papery bark. A few fall leaves, yellow and faded, still cling to their branches.

A mile farther, and the defile we've been following pours into a broad canyon on the far side of which stands a mountain clothed entirely in trees, a birch forest of tall straight trunks and spreading branches. There may even be pines among them. The mountainside, shaggy and green, presents a wall of welcome color. Fatigue from the day's march falls away. In our month on the moon I did not consciously long for the company of trees, but now, returned to their midst, I feel the joy of arrival, the relief of reaching a welcoming, almost familiar land.

50

Snow Leopard

WE LINGER among the birches. The exertions of Kang La have tired humans and pack stock alike, and with Bhijer cut from our journey, we have a day to burn. We spend it resting in a grove beside a shallow, sparkling river. A second reason for our layover is the predicament of Vishnu, the smallest and youngest among us.

He claims to be fourteen but looks ten. I never see him without his striped wool cap pulled down to his eyebrows. The cap has lop-eared flaps, from which the tie-cords dangle to his chest. His face is dark, his eyes quick and wary. You could not fail to notice him, and not just because he is so small among the adults. His profile would have done well upon a coin or a movie poster, and he carries himself with a lightness that catches the eye. His older brother had been asked to join the expedition, charged with bringing several horses, but an obstacle arose and the brother could not come. The horses, however, were still needed, as was the money to be earned by their rental. And so the family sent Vishnu in his brother's stead, as custodian for the stock.

Vishnu's cousins among the crew tease him much but also look out for him. He has proved an able worker, nimble and uncomplaining on the trail, and energetic in camp. All went well for him until yesterday at Kang La. After our wind-sheltered lunch below the pass, Vishnu and the other wranglers gathered the stock to resume the descent. But a horse in Vishnu's string was missing. He searched the vale where we had paused, looking into alcoves and behind boulders, but the horse was not there. Don't worry, his mentors told him. The animal has already gone down the trail, seeking better graze. We will overtake it.

But we didn't. And no horse tracks were seen once we exited the rockscape of the gorge and could read the sands of flatter country. The horse was not ahead of us.

The riding and pack stock grazed that night not far from camp, and Vishnu kept vigil over them until long after dark, hoping that the missing horse would straggle in.

But it never did. This morning, before the sky was light, he set out. He had no choice but to retrace our journey of the previous day, up through the stony gorge, even to the heights of Kang La. He must have worried for the safety of the horse, which in its solitude might have fallen prey to wolves or a snow leopard, but he probably feared the wrath of his brother even more.

While Vishnu labored up the pass, making a journey unthinkable for a boy his age in a more cushioned society, the rest of us tended to laundry, journal writing, or nothing at all. Our camp was spread through a wood below the mouth of the torrent we'd followed down from the pass. The torrent joined a tributary to Lake Phoksundo, which flowed over a broad, cobbled bed. Sandbars divided the braided channels, and on one of these Tonio, a genial outdoor guide from Alaska and one of our strongest walkers, discovered the paw prints of a large predator.

The tracks cross two patches of sand. The best imprint shows where the creature planted its hind foot atop the track of its larger forepaw. The prints are almost round, not oval like that of a canid—a dog, jackal, or wolf. A canid track would show an apex in the curve of the toes. The arc of these tracks is even. The pressed sand also conspicuously lacks the register of toenails: dogs and their relatives cannot retract their claws; cats can, keeping them sharp.

The forepaw tracks are four inches long, front to back: the cat was big. This was a snow leopard. Our Nepali companions say it was likely not the print of a full-grown leopard, which can be larger still, but easily that of a juvenile, a youngster on the prowl. There is no telling when the tracks were laid down, but it likely occurred not long before our little city of noise and movement arrived among the birches.

In the course of the morning nearly everyone troubles to hop the river braids to inspect the leopard tracks on the sandbar. To prevent an inadvertent footstep from destroying them and to make them easier to find, Tonio surrounded the tracks with rings of water-smooth stones, which rendered them curiously shrine-like. I visit the tracks with Wangmo, and again with Dr. Sonam and Amchi Lhundup, who have become accustomed to my questions about wildlife. The sight of the tracks now prompts greater volubility. Both men tell of attacks on livestock in their home villages in which snow leopards, singly or in pairs, killed sheep and goats, sometimes yaks. Lhundup reports one instance near his family's home in Mustang in which eighty animals were slain, producing such a surplus of hastily butchered meat that only a fraction of it could be consumed or sold, much of the rest going to waste, producing an economic catastrophe for the families involved. Wangmo mentions that the worst depredations seem to occur in the aftermath of very snowy winters, when the bharal have "moved away," or more likely died off. Springtime would find the leopards starving. And so the cats stalk the livestock of the villages, sometimes entering the actual villages and making their attacks in the stone pens adjacent to the houses. Until she went to school in Kathmandu, Wangmo had herself been a herder, and she says that when she took her charges out to graze, she was careful to keep them in the open and to avoid boulder-strewn ground where a leopard might lie in ambush.

A particularly gruesome detail, widely reported, is that snow leopards "drink the blood" of their victims. The misperception is understandable. Snow leopards kill sheep or goats with a throat grip, crushing the trachea and suffocating them. (They attack larger prey, like a young yak or, potentially, Vishnu's errant horse, at the nape, their teeth seeking the gaps between vertebrae and, through them, the spinal cord.) In the mayhem of an attack, with bodies thrashing, a throat on which a snow leopard had clamped its powerful jaws might tear, and blood spew. A witness, charging into the sheep pen with cudgel in hand, or a later visitor to the stained killing ground might think that a meal of blood had been the leopard's chief desire. Then, too, the aggrieved owners of murdered livestock naturally vilify their nemesis. Attributing a vampire's thirst to a varmint that slays more than it can eat and apparently revels in the act of killing makes a kind of sense.

From that point of view, the videos that Sonam, Lhundup, and I watch on Pau's smart phone make sense too. Pau got them from the proprietress of a house in Shey where he and the other two had gone for butter tea. The videos had been recorded months earlier during yarza gunbu season in high country east of Kang La. The first shows a single snow leopard, barely alive. Someone (out of frame) teases it with a haunch of meat, probably yak, while someone else yanks its tail. The second video shows two snow leopards, the larger surely the same animal as before, being roughly dragged by their tails.

The leopards, both juveniles, are lethargic, their hindquarters limp. Although one may have been slightly wounded on a foreleg and the other has a slack noose around its neck, no evident injury accounted for their passivity. Surely the cats had been poisoned. The second video ends as both leopards are slung like so much garbage into a cleft of rocks.

Estimates of the total world population of snow leopards hover near four thousand. The crucial question for their survival is whether adequate habitat can be reserved for them, safe from attacks by humans and safe also from the inroads of human livestock, which consume the herbage on which the leopards' wild prey depend. Unfortunately for the leopards, yarza gunbu has become a vital source of income for many Himalayan communities, an economic savior that partly counterbalances the widespread decline of agriculture, to which climate change, in turn, contributes mightily. The bizarre fungus draws thousands of local people and even more outsiders into habitats that would otherwise see only the occasional nomadic herder, or no one at all. Many of the yarza hunters come with pack animals. Their impact on the fragile tundra is heavy, and if they have a rifle and a bharal comes within range, *pow*, they collect the makings of a feast. Meanwhile, grazing competition between bharal and livestock of all kinds continues to increase in most areas, or at least remains high, assuring that snow leopards feel the squeeze of Petri Earth.

For the farmers and herders of Dolpo the snow leopard is understandably an enemy, a threat to their precarious existence, but they acknowledge that the rest of the world has a different view. The woman who shared her videos with Pau did so on condition that they not be divulged to the authorities. The rules of Shey-Phoksundo National Park are clear: snow leopards and other wildlife are to be protected. But the 1,373-square-mile park is not a park in the sense familiar to most Americans. Some nine thousand people live within its boundaries, and the enforcement of its rules is notoriously flaccid. Shey-Phoksundo is administered from a trim masonry complex beside the Suli Gad at the park's far southern boundary, and rangers rarely trek farther afield than Ringmo. Still, a video showing the torture and killing of one of the most exotic predators on Earth might prompt additional, bothersome patrols. Better to keep things quiet.

Geneticists tell us that the snow leopard diverged from its closest relative, the tiger (not the common leopard), about two million years ago, when the Himalaya was half its present height and our Sapiens ancestors still lived in trees. For at least the last million years, the cat has coevolved with the bharal, its favored prey. Their fatal dance is like that between the cheetah and the gazelle, which has engendered in both species the ability to sprint at dazzling speed. In the case of snow leopards and bharal, their ages-long duel has produced an escalation of stealth, as well as detection, in tandem with the capacity to endure one of the most punishing environments on the planet.

Zoos will preserve many splendid and fascinating creatures deep into the future but they cannot deliver the relationships that formed them. Only wild environments can do that, and wild environments, squeezed by Petri Earth, are everywhere embattled. A recent report asserts that "around 1 million species already face extinction, many within decades." Those million species constitute about a quarter of all the animals about which enough is known to support an assessment. The snow leopard is among them, and although the bharal is fairly common today, the time may come when it, too, joins the ranks of the threatened.

We know that things will get worse before they get better. The momentum of climate change guarantees it, to say nothing of the assured increase of human population by several billions more. Such prospects behoove us to become latter-day Noahs. We need to build arks, vehicles to carry the beauty and diversity of Earth's present creation across the inhospitable seas that lie ahead. Some of us, like George Schaller, have been at it already for decades, building arks in the form of parks, wildlife refuges, reserved forests, wildlife corridors, restricted use zones—the potential designs are legion. Old, leaky arks, like Shey-Phoksundo National Park, need an overhaul to become seaworthy, and even then, the voyage is sure to be rough. The imperative is to launch as many such arks as possible. Some will sink, some will be blown off course, and not all the rest will get through with their cargoes intact. But a hopeful spirit, committed to intrinsic good and with faith in surprise, compels us to make the effort. We cannot know the outcome. If we prepare, not for the worst, but for the best, some of our arks will complete their voyage to better days. The more we build, the more will make it.

Such an effort will require the kind of determination Vishnu showed as he searched for his brother's horse. Through the long climb up the defile to Kang La, he saw nothing of the animal. Nor did he find it where we paused for lunch, in the wind-sheltered bowl where he first detected its absence. Vishnu kept climbing, up and up, with the pea gravel sliding underfoot, requiring him to double his exertions. He continued to the very top of Kang La. From there, he scanned the great snow-streaked cirque. He looked and looked, examining every slope and gully. Finally he thought he spotted a shape near Black Pond. It was the merest speck but it seemed not to be boulder. Then the speck seemed to move. Yes, it moved. He was certain now. At once desperate and exhilarated, Vishnu flew down through the switchbacks, down the trough between snowbanks and across the boulder fields to the lake. The speck indeed proved to be his truant horse. It neighed in recognition. It was hungry, scared, and lost, and it was every bit as glad to see the boy as the boy was to see it.

Vishnu's victorious return makes the camp merry. His comrades build bonfires of birch wood, and the stories and jokes begin to flow. Someone produces a drum and begins to sing. Others join in. They celebrate Vishnu in long, repetitive songs, improvised on the spot, with teasing, limerick-like verses. Well before the embers of the fires cease smoldering, Vishnu is sleeping, I presume soundly. Eventually I sleep well too, but before drifting off, I lie some minutes in my tent, eyes closed but still seeing the videos of the poisoned snow leopards, the embers of their eyes also smoldering.

51

Amchi

THE EVENING of Vishnu's return, Amchi Lhundup spoke to the group about the traditional medicine of the Himalaya. In a crowd, you would not remark Lhundup. He is of medium height and medium build, unflashy in dress, and possessed of ordinary good looks. At a distance, he does not stand out. But close up, you feel the attraction of his kindness. He is thirtyish, quick to laugh, and possessed of a warmth that anyone would be pleased to sit beside.

Lhundup began his talk as though it were a formal lecture, and he was nervous. His pages of notes, the cramped script scarcely visible in the falling darkness, trembled in his hands. Our medical director, Charlie McDonald, had invited Lhundup's presentation and mumbled to me his remorse: he had not meant for the occasion to entail so much preparation or pressure.

But Lhundup's audience was rapt, for he had made a strong impression on the other medicos. They had observed his close rapport with his patients, the way he touched them and held their arms for long minutes as he listened to their pulse, or as he would put it, their pulses, for he felt for distinct heartbeats in different places, the better to unlock the secrets of organs and tissues. He sat close to his patients, smelled their breath and the odors of their bodies, assessed the color of their skin and felt its moisture, tension, and temperature. Above all, he looked into his patients' eyes, with a gaze neither aggressive nor sentimental but open and clear. His intent was to go past the physical surface and into the person. And his patients would say he succeeded. In being treated by Lhundup, a person felt that he or she had been deeply read, like a book pored over by a scholar.

Charlie said Lhundup's father, Amchi Gyatso, embodied the amchi way. On an earlier Nomads trip, Charlie watched Gyatso closely: "There was something that he had, that I didn't have, which was an ability to connect in milliseconds with the patients. He would walk in, he would be *with* them, and only them, and he would touch them, and interact with them in a way that, no matter what he did, whether he gave advice, or whether he gave them amchi medicines, they walked out feeling better." It was a brand of healing untaught in American medical schools, consisting not so much of therapeutic procedures but of something closer to a laying on of hands. Charlie said that observing Amchi Gyatso inspired him to revise entirely his approach to his patients.

Bit by bit, Lhundup conquered his nervousness. With the river purring in the background, he told us that the roots of amchi medicine reach back more than two thousand years and a thousand miles into the steppes of Mongolia. From there the skills traveled into China, Tibet, and eventually the Himalaya. Amchi medicine recognizes 404 distinct diseases. Of these, 101 can be cured with foods and remedies drawn from the herbal and animal pharmacology of the region. Relatively speaking, that's the easy part. Another 202 diseases, owing to the effects of past-life experiences, require spiritual as well as medical treatment. This realm of healing involves, not just diffusions, powders, poultices, and massage, but pujas intended to restore balance in mind and soul. An amchi frequently assists in conducting such ceremonies, usually in cooperation with lamas or other spiritual guides.

The final 101 diseases, a quarter of the universe of afflictions, are considered immune to the amchi's powers, said Lhundup. They involve past-life problems so great as to make them incurable. Medicine, even spiritual medicine, has limits.

Traditionally amchi medicine was passed down solely from fathers to sons, but many chains of transmission perished with the untimely death of a father or the absence of a male heir. Lhundup's late grandfather dreamed of founding a school in his home city, the ancient capital of Lo Manthang, in upper Mustang. He hoped to institutionalize amchi knowledge, freeing it from the limits of patrilineal transmission and allowing it to reach—and serve—a wider world. Lhundup's father and uncle, sons of the grandfather, realized that dream when they founded such a school in 2000, not in Lo Manthang, but in the bustling city of Pokhara in central Nepal. Further, they took the radical step of accepting women in their curriculum, Lhundup's sister being among the first female students.

In amchi medicine a belief in reincarnation underlies the doctor-patient relationship. According to Lhundup, "The amchi must treat every patient as though she is his mother or he is his father, because, in a previous life, possibly this was the case." A skeptic may find the rationale dubious, but such a notion yields a depth of compassion that any patient would welcome.

I took solace in what I learned from Lhundup and began to see past lives in a new light. Evolution is past lives. Our heritage of DNA is the result. From it, we inherit our selfish genes and our run-amok drive to colonize and exploit every habitat. Imagine, for a moment, that the dilemma of Petri Earth results from a disease shaped by past lives. Possibly Sapiens *had* to arrive in this predicament. The hard wiring of our genes decreed that it be so. While such a view can lead to resignation and inaction, it can also be the means for reconciling with the historical moment—for finally not finding fault with the present and for freeing one's energies to get on with the work of ark building, delivering care, or following one's vocation wherever it leads. Perhaps humanity's past-life problems belong not to the realm of the incurable but to that of the merely difficult. This is a question without an answer, a conundrum that our compendious knowledge of the cosmos cannot solve. We are left in a state of not-knowing, the state in which hope abides.

52

Yeti

ONE MORE thing I learned from Amchi Lhundup was how to defend myself from a yeti.

Tales of wild, semi-human creatures are told around campfires the world over. Years ago in the forests of central Laos, the porters and guides I traveled with spoke of a red and hairy little man called the phi kong koi, who was no less fearsome for being small. His name derived from his terrifying nighttime call, which one of the guides claimed to have heard in the course of our journey. Other guides attested that they had heard the phi kong koi on other occasions, or if they had not, they knew someone who had. Another quasi-human bogeyman, the Sasquatch, is said to dwell in the Pacific Northwest of North America, and additional ape-humanoids appear in the folklores of every inhabited continent. Some scholars say that such creatures express a human compulsion to conjure a mirror being that is wilder and freer than ourselves. Others say they represent a deep memory of days long past when Sapiens shared more of the world with other large primates and even other hominids. Still others say that some of those mythic beings are real. In *The Snow Leopard*, Matthiessen reports Schaller and himself as at least agnostic on the existence of the yeti, and in a momentary encounter on the upper Suli Gad, when something reddish leapt with peculiar agility behind a boulder, Matthiessen wondered if he might have glimpsed one.

In a gompa along our route, a monk showed us an aged, painted cloth bearing depictions of the animals of the region. There was a horse, a dog, a tiger, a coiled serpent, an owl, and a fanciful, smiling fish. There was a bearded goatlike creature, with rearward pointing horns, as well as several deer, with spike antlers pointing forward. There was also a female yeti. She had wide, fearless eyes, a head of unruly hair, and shorter hair all over her body, including her long-nippled breasts. Her sexuality must have impressed the artist mightily, for her pudenda were so large and forward-projecting as to be threatening. There could be no doubt that she was powerful, as wild as a wolf, and terrifying. No lone traveler would desire to meet her.

Lhundup explained that yetis are more common in Mustang and Humla than they are in Dolpo. He did not say why. Still, yetis can appear almost anywhere. He told me the most important thing to remember is that, if you meet one, you must not run uphill. A yeti is far stronger and more agile than any human. Before you can think two thoughts, it will bound up the slope and snatch you. And then possibly eat you. The thing to do, said Lhundup, is to run downhill. That's your best bet. As the painting in the gompa showed, the hair of a yeti is long and disheveled. When the yeti, be it male or female, plunges downhill after you, its hair will fall like a curtain in front of its eyes. Thus blinded, the yeti will become clumsy and slow. Perhaps it will trip, even fall. This will give you a chance to get away.

With sincerity, I thanked Lhundup for this knowledge, which I hold as dear as anything I learned in Dolpo. It was not so much the lesson that I treasured as the kindness with which it was taught.

53

Clinic Notes

NO ONE gets out of here unchanged. The lessons of the clinics are never far out of mind, and the agents of change, more often than not, are the patients. At one clinic a paralyzed young man crabbed sideways past the triage table into our cluster of tents. His mother walked beside him. He moved horizontally to the ground, with one unbending leg stuck out as stiffly as a tree limb and the other cramped beneath his body. His calloused fists punched the earth in alternation so that he swiveled slowly across the barley stubble. Prem assigned him to Michael Lobatz. One of our guides, Dawa Sherpa, a mountaineer who has summited K2 and Everest, led him to the specialists' tent and waited with him.

The young man's predicament recalled a passage from *The Snow Leopard*. Heading west from Pokhara toward Dolpo, past the last of the roads then existing, Matthiessen and Schaller passed a girl "dragging bent useless legs." Perhaps in his description of the girl, Matthiessen had in mind to foreshadow the crippled lama he would meet at Tsakang. Perhaps he meant only to record what he saw and what moved him. His words have moved many readers since: "Nose to the stones, goat dung, and muddy trickles, she pulls herself along like a broken cricket." The sentence stops me cold. No phrase better evokes the cruelty of the world than "broken cricket." Yet the image also sets the girl at a distance, away from the reader, which is the way most of us witness grave misfortune.

At the clinic, the young man with the pretzeled body, broken though he was, did not seem broken. His gaze was warm, his eyes clear. Later, when we met in circle, Dawa said that when she looked at him, he did not flinch but looked back openly and softly, revealing a deep calm. Dawa is no stranger to sacrifice and trial, and she said that the gaze of the paralyzed young man moved her deeply: "There was a lot there, a lot of beauty."

Michael Lobatz also spoke. The young man, he said, was about twenty. He had suffered a stroke within months of birth, and his paralysis and deformity were the result. Michael thought that, with injections of Botox to release frozen muscles, together with a program of physical and occupational therapy, the young man might literally get on his feet again. "This boy," he said, "probably has normal intelligence and could have a reasonably normal life." Michael's voice tightened, as he labored to say, "He could be more than a crawling person on the ground." The words came out high-pitched, strangled. We waited a long moment for Michael to gather himself. "Arrangements could be made in Kathmandu," he said, his voice regaining its tone.

Michael explained that he had discussed the possibilities with Dr. Sonam and with the young man and his mother, who solemnly thought the matter over. Lobatz admitted being stunned when the mother politely declined, but her choice was clear and her son seemed wholly in agreement. There was no reasonable way for him to "be brought down" from the heights of Dolpo to a road where transport might be procured. Even for the able-bodied, such a journey entailed days of walking. And the medical interventions, the mother foresaw, would transform her relationship with her son and alter the balance of their lives together, which they cherished without resentment. No one had to explain that when the mother died, perhaps not long in the future, the son would die too, for there would be no one to take care of him. That was simply how things were.

Lobatz did not dispute their choice. Still, he felt immense frustration. We had talked in previous meetings about "moral distress." The paralyzed young man with beautiful eyes made such distress palpable and immediate. Roshi cautioned us on the need to recognize limits, and not just of what is attainable. Not every "cure," she reminded, is compassionate. Not every "improvement" reduces suffering. Citing another patient whom we treated that day, she raised the danger of "pathological altruism," a topic on which she frequently writes and talks. Altruism, she explains, is an "edge state," like empathy or grief. It begins by helpfully opening the heart but, taken too far, has the potential to overwhelm the caregiver. The urge to help can take a person past the edge of balance, silencing other judgments and instincts. The person becomes subordinate to the urge. Hence burnout. Hence depersonalization, futility, and despair.

The patient who prompted these thoughts was a weeks-old baby girl, her face a mass of leaking sores. The impetigo so disfigured her that she could neither see nor nurse. She had clubfeet and a cleft palate and was badly underweight. Even were she able to nurse, her mother produced no milk. The mother sat, hands in her lap and a thousand-yard stare in her eyes, as the medicos examined her baby. When Wangmo asked other nursing mothers in the village if they might spare milk for the stricken child, they said, "What's the point? She will die. And if she lives, what then?" Our doctors treated the baby's impetigo with creams, while ruing our pharmacy's lack of liquid medicines that might have served better. The mother, forty-one, was dehydrated and likely had type 2 diabetes and one or more STDs. They gave her liquids and vitamins. That, plus counseling, marked the limit of our clinicians' powers. The mother would bring the baby back tomorrow, and we would see if the creams had helped. (Indeed they did help, and probably saved the baby's life, but the clinicians who rendered this act of mercy continue to question if it was indeed merciful. A year later the mother brought the baby to Kathmandu to seek help from Wangmo. The baby had scarcely grown. She was severely microcephalic: her brain had not and would not develop normally. There was no serious prospect that her profound mental and physical disabilities would improve.)

In the afternoon following that particular clinic we sat on stools in a circle beside the tents, a chill settling like dew with the waning of the sun. As ever the little black crows called choughs spun and squabbled overhead. The reports and observations continued. There was talk of malnutrition among pregnant and nursing mothers and among the very young. Mild distensions of the belly had been noted in a few cases, but subtler problems were no doubt widespread. Dr. Sonam said that deficiencies in vitamins A and C and possibly other nutrients were probably typical for children of the region.

The conversation shifted to diagnostic quandaries provoked by our patients' other ailments, the pervasive joint pain, dizziness, gastritis, general weakness, and recurring migraines. There was laughter when someone pointed out that Prem and the triage nurses must be losing focus: they had long since taken to writing "whole body" in the space for "complaint" on a majority of intake forms. There were expressions of thanks, Wangmo's in particular, who said that without the kindness of her grandmother and others she would be trapped in the same toilsome life endured by her patients in the women's tent. And there were tears too, including Wendy's, when she reflected on what those women "have to go through," the drudgery, the continuous pregnancies, the shunning associated with menstruation. And there was satisfaction as well, and not just from the provision of analgesics. Michael Lobatz described a mother's relief when he told her that her child's migraines would get better. He said doctors are never more fortunate than when they can provide the gift of hope.

The differences between patients in Dolpo and patients in the US attracted frequent comment. One doc said that most of his stateside patients were what he called the "worried well," people whose anxieties outweighed their physical problems. "Don't get me wrong," he said, "a lot of people need intervention, and to provide that is very satisfying, but the practice is not like here where you think at least that you are having more impact." His friend and colleague from Colorado said that it was the gratitude and resilience of the Dolpo-pa that impressed him most: "They come in with joint pain, a blown-out knee, GI distress, a horrible rash, whatever, and maybe we can't help them. 'Sorry,' we say. 'Wish we could do something for you.' And they get up and smile. They say, 'That's fine. Thanks anyway.' And off they go, as cheerfully as they came in. Patients back in my clinic are so different. Whatever hurts them becomes so much bigger a thing. And we give them meds for blood pressure or pain but they really seem to want us to fix something bigger than that, something we don't have meds for. They want us to fix the pain that is in their minds or in their souls. My Nepali patients have lots of problems, but not that one."

54

Kanjiroba

THE RIVER that will lead us to Lake Phoksundo glimmers in early light, braiding among sandbars, each bend and riffle refracting the light in its own way, rays dancing on the water. Of all our mornings, this may be the most beautiful. Ragtag elements of the expedition converge toward the main trail: a clutch of mules, loads swaying, navigates a copse of red-stemmed willows; horses and riders trot in single file past a fall of yellow boulders; a quartet of porters bearing awkward baskets hustles up a shadowed bank. The warming air fills with the low music of hoofs and harness, footsteps and birdsong. Hikers in twos and threes join the procession, the vivid colors of their jackets brilliant against a dark cliff. Something has changed. A liveliness animates the movement of animals and people. A new spirit lifts us, and it is not just the bright day or the recent rest. We are horses headed for the barn. The major clinics, the ancient, remote gompas, the lunar heights, and the toil of the great passes lie behind us. We have a week of walking yet to go, but a kind of countdown has begun, toward paydays for the crew and hotel comforts for the Westerners, and toward baths, clean clothes, and homeward travel for everyone.

Still, we have one more mountain to meet.

From the tallest passes we glimpsed a cloud-wreathed range, towering in the west from Bogu La and in the south from Kang La, always distant yet present, stark and shining when the wind drove off the clouds, the jagged summits shedding plumes of wind-driven snow that feathered far into the sky. This was Kanjiroba, a massif cresting just shy of twenty-two thousand feet, a height taller than the highest points of Europe, Africa, and North America. The mountain now looms at our right hand, and for the next two days, we will skirt its eastern foot. We will drink its snowmelt water, breathe air chilled by its glaciers, and sleep in its broad shadow. We will feel its immensity, as the largest thing in a land of large things, and wonder at the alien, ice-girt world of its summit.

Nearly all the ranges that comprise the Himalaya follow the arc of collision between the Indian and Asian plates, which runs roughly east and west. But the Kanjiroba range stretches north and south, perpendicular to the arc. It traverses the zone of collision, linking the Himadri, the "Great Himalaya," home of the tallest peaks, to the rain-shadowed deserts of the Tethys Himalaya, which includes Upper Dolpo. How this came to be, how the crumpling of Earth made a right-angled turn in this particular place, adds a scientific puzzle to Kanjiroba's inherent mystery. It is a geologic anomaly, a reminder of how much we don't know.

For as long as we travel beside Kanjiroba, we never see the peak. We see ramparts, walls, and skyscraping ridges, but the arctic summit remains shielded by the bulk below it. Late on the first morning, we rest on a bluff beside the river. The bluff results from an ancient landslide that crashed down toward Kanjiroba from heights to the east. Immense boulders, some canted at puzzling angles, lie scattered across the top of the bluff. One of these, near the center, came to rest facing the sky, a natural table, almost a stage, smooth and spacious. On it someone—or rather a crew of someones—has placed a large, squarish block of limestone bearing mantras both ancient and indecipherable. The etched letters of the mantras are filled with red pigment and flow across the stone in tight garlands of script. Scratched among the largest words are smaller messages, written now horizontally, now vertically, as though in annotation. Passersby have piled colored stones and other offerings behind and beside the limestone, but the space in front belongs to the curved horns and weathered skull of an argali, a wild sheep now extirpated in these parts, set before the mantras like a sentry.

We lounge in this place, gazing across the river at the flank of Kanjiroba, as countless travelers before us, including those who etched the boulder, must also have done. The snow-hung wall of the mountain is fluted with avalanche chutes that lead to glaciers at their foot. Prem is beside me. He says that twenty years ago, when he first traveled this route, the glaciers descended twice the distance they do now. He points, "All the way to there," indicating the tops of two gullies that look strangely out of place, abandoned by whatever force had cut them. Between the tops of the gullies and the current bottom of the glacier, the rocks are smooth, the earth mostly bare, the vegetation scant and hesitant. The ground has the look of a half-healed wound.

It is well known that the glaciers of the Himalaya, like all the frozen landscapes of the planet, are shrinking, but the manner of their recession is complex. Some glaciers shrink fast, some slowly, some not at all. A few even seem to grow. But measurements of the overall trend put the average loss at seven inches of depth across the whole extent of ice, every year. A study that focused on the Nepal-Sikkim region calculated that approximately one-half of one percent of ice mass was lost every year from 1962 to 2006. Such attrition would account for a depletion of more than one-fifth of preexisting ice. Other research suggests that the Himalaya's higher elevations are warming nearly twice as fast as their foothills and up to three times the rate of the rest of Earth. Some believe these differences are increasing. The reasons for high-altitude landscapes' warming especially fast include the diminishment of albedo—the reflectance of the land's surface. Kanjiroba's bare lower slopes offer a good example of this. Without bright snow and ice to reflect sunlight back into space, the comparatively dark, newly bare ground absorbs and retains more heat. This, in turn, pushes the snowline higher, in a self-reinforcing cycle.

The people of Dolpo have noticed such changes, and they explain them in interesting ways. Some of the elders say that the receding glaciers and unsettled weather of recent years are the result of outsiders coming to pick yarza gunbu. Not knowing the land or its ways, the outsiders break unwritten laws, abuse pastureland, pollute streams, and cut shrubs and trees where no cutting should be done. Such behavior upsets the spirits of place. As a result, brutal winters now alternate with winters that are hardly cold, and avalanches fall where avalanches never fell before. The rains also seem to be affected. Sometimes they start too early or end too late. Or they don't come at all. And the rituals that formerly restored order when things slipped out of balance are less reliable than they used to be. They seem unable to overcome so high a level of disturbance.

The Dolpo-pa did not create the bundle of woes that melts their glaciers and renders their grasp on subsistence more tenuous. In Dolpo, per capita output of atmospheric carbon is negligible. Like poor people everywhere, they bear the brunt of the developed world's profligacy, which shows little sign of abating. Their consolation prize in recent years has been that they, too, participate in the yarza gunbu gold rush. Their share of the proceeds has generated a minor economic boom in Upper Dolpo, which offsets other losses, at least for now. Meanwhile, the young people of Dolpo know very well that the big cities hold such opportunities as they might hope to find, and so off they go to Pokhara and pollution-choked Kathmandu, leaving the drudgery of the barley fields to aging parents and grandparents.

Almost every corner of the world is caught in the same gyre of change. I am no world traveler, but I have burned more than my share of jet fuel, and the ubiquity of transformation, even in the remotest places, is striking. Motorcycles and packaged food in the Lao jungle. Satellite TV in the Arctic. Cell phones in the Amazon. Modernization has leapfrogged into the last blank spots of the global map. In some places the pace of change is rapid and noisy, in others nearly invisible, but the penetration of cash and consumer goods creeps forward like a rising tide.

I am dozing when Prem stirs and says it is time to go. Stiffly I get up and look anew at the sawtooth dazzle of Kanjiroba against the sky, and then at the mountain's fluted wall and the gleam of graying ice. Below the ice, the land is bare, brown, and scraggled with plants. The Himalaya, once the abode of perpetual snow, now approaches an end to that particular species of perpetuity.

Late in life, the American novelist F. Scott Fitzgerald had a breakdown. Alcoholism, nervous exhaustion, family tragedy, writer's block, and penury played their part—scarcely a woe failed to join the gang of ills besetting him. When finally he wrote about his "crack-up," he said his ultimate undoing was that he'd lost "the ability to hold two opposed ideas in the mind at the same time, and still retain the ability to function." Kanjiroba contains the kind of contradiction Fitzgerald was talking about. It is no less sad than it is beautiful. It is a glorious disaster, a magnificence of loss. I can hold these feuding thoughts in my mind—they scrape against each other, but they can coexist. So far, so good. The bigger problem is the pair of ideas that Fitzgerald said finally broke him: "That one should, for example, be able to see that things are hopeless and yet be determined to make them otherwise."

55

Cutting Deals

IN A grove of pines Jigme, Nurbu, Rigdzin, and a few others have stopped to rest. I join them. By this point in the journey, all of us feel the toll of the miles and we are empty of small talk. A few snacks are passed around. Then we arrange our packs as pillows, lie back in the wiry grass, and stare into the green canopy of the trees. Spangles of sky shine through the branches. A low chirr of insects undergirds the quiet. Far off, a lone bird peeps a single, metronomic note. Though the air hardly stirs, the needles of the pines murmur gently.

The forest is busy even when it is not busy. Insects hatch, crawl, and forage. Bacteria and fungi, the refineries of the soil, convert organic debris into hundreds of chemicals, which trees and other plants take up and move around, root to root, linked by webs of fungal mycelia as intricate as the circuitry of a brain. Organisms large and small twine the forest into a network of interdependence. Birds and worms, beetles and spiders, moths and rodents, the unseen invertebrate no less than the conspicuous mammal, take and give in uncountable ways, producing a hive of exchange, a continuously negotiated economy of cooperation and competition, out of which every discharge is recycled for new consumption. There seems to be a kind of altruism in this turbulent metabolism: sometimes trees shunt nutrients to the weakest members of their stand, or they release pheromones to warn of insect attack so that the downwind forest can muster a chemical defense. It even seems that, when death is imminent, some trees will flush nutrients and other valuable compounds through their roots to make a bounty for their neighbors, lest the chemicals decay unused. In Richard Powers's sweeping ecological novel *The Overstory*, a character says of the forest she studies, "Everything out here is cutting deals with everything else."

The deal making of the natural world is one of the great post-Darwin discoveries of biological science. For Darwin, evolution proceeded species by species. Such a view reflected his moment in history. He had to start somewhere, and the predominant view of Western culture in his age was that competition among individuals produced the outcomes of life. Human agents of change—such alpha males as statesmen, industrialists, and generals—appeared to be singular and independent, and it was easy to assume that the driving forces of the natural world were similar. Much harder to see were the interconnections among species, which scientists have since teased out through observation and analysis. We now better comprehend that change rarely, if ever, occurs in isolation and that the true unit of evolution is less the species than the ecosystem— whole communities bound together in relations of reciprocity. To a large degree, it is the ecosystem that collectively accepts, amends, or rejects the changes that arise within it.

Sometimes in the literature of science, a writer shouts almost gleefully, "Darwin was wrong!" while implying, "And I am right!" How brilliant. Clearly Darwin did not grasp all of biology in a single lifetime. What he provided was a framework, previously unseen, that his posterity continues to flesh out and revise. A case in point: Darwin's uniformitarianism, borrowed from Hutton and Lyell, was more accurate than the catastrophism that it replaced, which assumed the validity of divinely engineered calamities like Noah's flood. But he could not have anticipated real catastrophes like the Chicxulub asteroid that slammed into Earth sixty-six million years ago, filling the atmosphere with enough dust and debris to disrupt planetary climate and cause the mass extinction of non-avian dinosaurs. A century passed after Darwin's day before that chapter in Earth's biography was even halfway understood.

Nor could Darwin, without the perspective of microbiology, have appreciated how indefinite the boundaries of the individual really are. My Buddhist friends speak of the ephemerality of the self, of the idea that the ego is imagined, not actual. Science adds to such ambiguity. Are the colonies of bacteria in our gut part of our "self," or are they separate from us? It turns out that such bacteria outnumber (but don't outweigh—human cells are larger) the cells that are properly our own. Meanwhile, the mitochondria that govern the release of energy in our cells pose an even more puzzling question. They turn out to be evolutionary hitchhikers that found a supportive habitat in the cytoplasm of our single-celled ancestors. They have their own genome and reproductive process, and they "pay the rent" for living in our cells by keeping us alive. But are they "us" or are they "them"?

Darwin lacked tools to probe the intimate exchanges by which organisms mutually benefit each other, and he lacked the lens to see that they evolve, not solely as species, but frequently in groups of species—or, looked at "more granularly," they evolve in concert with innumerable other organisms within their ecosystems. Yet he firmly grasped that not all of the struggle for survival was competitive. He understood that "fitness" might also embrace altruism, cooperation, and even love. In *The Descent of Man* he discusses at length the "mental powers" of non-human animals, concluding not only that many of them possess emotions like ours but that their intuitions comprehend a "moral sense" that gives rise to acts of fidelity, kindness, and service. Darwin concludes that "the difference in mind" between humans and the higher animals, "great as it is, certainly is one of degree and not of kind."

A case in point, one of scores of examples in *The Descent of Man*, involves a keeper of animals at the London Zoological Gardens. The keeper was kneeling on the floor of an enclosure when a large baboon attacked him, biting and tearing at the nape of the keeper's neck. Also in the enclosure was a small South American monkey "who was a warm friend of this keeper." The little monkey was "dreadfully afraid of the great baboon," but, seeing "his friend in peril, he rushed to the rescue, and by screams and bites so distracted the baboon that the man was able to escape, after, as the surgeon thought, running great risk of his life." Darwin himself inspected the keeper's "deep and scarcely healed wounds." The facts of the incident were indisputable and provided yet another instance of a "lower animal" exhibiting bravery and selflessness, qualities confirming the creature's possession of "a moral sense."

Many of Darwin's successors denied the existence of higher faculties in nonhuman animals, much as they also ignored his emphasis on sexual selection. In the first instance, they insisted on seeing animals as yes-no robots responding to positive or negative stimuli. This mid-twentieth-century view suited the experimental reductionism that biologists, striving to emulate the "hard sciences," borrowed from physics and chemistry. In the second instance, the patriarchy of Darwin's era found it humiliating to accept that females, not males, should have made the decisions that shaped the evolution of so many species, including our own.

Science gets many things wrong, but over time, it tends to self-correct. Ultimately, the better, truer story wins out. Darwin was of his time but also ahead of it. Part of his gift was to see connection where others saw none, most importantly between humans and the rest of the animal world. It is arguable that a knack for appreciating

connection—versus resistance to doing so—is among the most deter-
minative features of the human psyche. A penchant for seeing connection, or the lack of such an inclination, helps to explain, for instance,
why climate change has become not just a subject of debate but the
object of unending dissension and vitriol.

Acceptance of climate change requires accepting a connection
between human behavior and the condition of the planet. That's simple
enough, although it goes against the interests of powerful economic
forces. The implications continue: if human behavior can affect the climate, it can affect anything and everything. Climate change invites—
no, it demands—an ecological view of the world, a view in which life
on Earth is a mass of entanglements where everything is connected
to everything else. Once connection is accepted, a worldview built
upon the primacy of the individual begins to wobble. For starters, if
the problems of the world are global, and if the people of the world are
connected by shared contributions to those problems and shared challenges in facing them, then it follows that they also share obligations to
each other. From there the connections rapidly grow thicker, so that the
notion of a special tribe, a superior race, a chosen people, or any other
kind of exceptionalism becomes difficult to sustain.

Conversely, a worldview of separation, in which maintaining a sense
of *us* requires a rigid definition of *them*, repudiates this web of connections. If my sense of self depends on what differentiates me from all
those *others*, then pulling even a single brick from the wall of separation
can threaten the collapse of much more. Identity, even cognitive structure, may be at risk. So climate change be damned. Don't tread on me!

Darwin, true to his abolitionist roots, understood the implications
of connection. "As . . . small tribes are united into larger communities,"
he wrote, "the simplest reason would tell each individual that he ought
to extend his social instincts and sympathies to all the members of the
same nation, though personally unknown to him. This point being
once reached, there is only an artificial barrier to prevent his sympathies
extending to the men of all nations and races."

But Darwin was hardly naive. He well understood how difficult
the embrace of connection might be for an individual inclined against
"others": "If, indeed, such men are separated from him by great differences in appearance or habits, experience unfortunately shews us how
long it is, before we look at them as our fellow-creatures."

56

Hospice

A CONSERVATIONIST friend argues that to prophesy the end of anything is an act of hubris. We cannot know the future, she says, and we delude ourselves if we think we can. To foretell the demise of great waves of the world's wildlife is an act of overweening pride. She may be right. Chance never sleeps. My own hope, such as it is, lies in surprise, and surprises of various kinds may ultimately alter our calculation of outcomes for the wild world. This may already be happening. Some of the contra-indicators for the Sixth Great Extinction include:

✦ Urbanization: as more people crowd into megacities, some rural areas become depopulated. Without constant hunting, wood-gathering, grazing by domestic livestock, and other human-caused disturbances, many natural systems, including their wildlife, will rebound. India provides examples of this, and in some of the vacated rural areas, wildlife—even tigers, provided they are protected from poaching—are making a comeback. (The trend, however, can also reverse: not just in India but throughout the world, the economic impact of the Covid-19 pandemic has sent many impoverished city-dwellers back to their natal villages and to a resumption of subsistence activities. The collapse of tourism, meanwhile, has deprived many villagers of the means for living lighter on the land.)

✦ Speciation: some authorities maintain that rapid adaptation and global mixing in wildlife populations is producing not a sixth extinction, but a sixth genesis. In a way, it is a failsafe argument: evolution guarantees that life on Earth, given a chance, will diversify. The question is: Will the rate of diversification surpass the rate of loss? Those who say yes seem to rely more on faith than data.

✦ Technology: perhaps scientists will manage to accelerate biological adaptation to new global conditions. One promising area involves developing new strains of coral capable of tolerating warmer and more acidic seas, as Madeleine van Oppen and the late Ruth Gates have striven to do at their respective labs in Australia and Hawaii. Perhaps such new strains will be successfully seeded into the Great Barrier Reef and other cornerstones of marine diversity, and perhaps analogous "fixes" may be found for other ecosystems. That's a lot of "perhapses," but, hey, don't be a party pooper.

✦ "Geoengineering," the application of technical fixes at a planetary scale, may avert the worst impacts of climate change, benefitting natural systems as well as people. So say the advocates of re-jiggering the planet. One proposal involves deploying space-tech umbrellas in the upper atmosphere to reflect more of the sun's energy away from Earth. Another calls for pumping vast amounts of water (with what energy?) back onto the Greenland and Antarctic icecaps to refreeze during winter, thereby forestalling the rising of the seas. Of geoengineering one thing is sure: its potential for moneymaking is limitless. As conditions worsen, desperate societies will clutch at increasingly wild ideas, while companies hungry for giant contracts will launch sales promotions at the scale of national political campaigns. Perhaps such schemes will live up to their hype. On the other hand, perhaps they will amount to no more than a lavishly expensive ghost dance for industrial capitalism.

✦ Enlightenment: humans, the hopeful say, will rein themselves in, protecting and restoring habitats on a massive scale. Maybe so, but so far, not so good. Protected areas, which often exist more on paper than in actuality, and the careful management of surrounding buffer lands have certainly slowed negative trends, but they are far from reversing them on a broad scale. And people, in general terms, continue to behave . . . well, like people.

✦ Epidemic: disease might decimate human population, as actually happened in the fourteenth century when the Black Death triggered a rewilding of many parts of Europe. A much greater calamity swept the Americas when Old World diseases, on the heels of Columbus's voyages, reached populations unadapted to them. (The consequences—because forests reclaimed vast amounts of abandoned farmland—may have included a lowering of atmospheric CO_2 and a cooling of the climate, producing the so-called Little Ice Age, which began in the sixteenth century.) A repeat experience, enacting tragedy at a colossal scale, is horrific to contemplate but lies within the universe of possibilities. The economic shutdowns brought on by the novel coronavirus in 2020 produced hints of nature's potential for resurgence: peccaries, emus, elephants, and other species were seen to be wandering city parks and streets (although many supposedly corroborating videos proved inauthentic). While such anomalies make for entertaining YouTube clips, beyond the roads and houses, back in the swamps and forests, genuinely significant extensions of habitat may also have occurred. Even if such changes prove transitory, they illustrate one way in which rewilding can start. Outcomes that are worse for humans and better for most critters may ensue with the inevitable onset of the next pandemic. But the price to pay may very well exceed our most ghoulish imaginings, and such a cost, like the cost of every terrible thing, will fall most heavily on the poorest and most vulnerable among us.

✦ Combinations of the above: unexpected linkages produce surprise. Although ignorant of the future, we can nevertheless prepare. If we build enough arks and keep them afloat, and if the current expansion of Petri Earth eventually abates, or at least shifts to a more forgiving phase, perhaps the worst outcomes now predicted may be averted. And then what? A new and enduring equilibrium? Maybe, so long as runaway warming does not take us past thresholds from which there is no return. But don't count on it.

Let's be real: we don't live in the gentle Holocene anymore. Alteration of the climate has delivered us to the Anthropocene, and the heat already loaded into the climate system guarantees increasing impacts for decades to come. Even if we start doing everything right tomorrow, our path will not lead back to where we used to be. As Bill McKibben has observed, "We've lost [the] fight, insofar as our goal was to preserve the world we were born into. That's not the world we live on any longer, and there's no use pretending otherwise."

We are in a sick room. The patient is a tough old coot, still with plenty of vigor, but key systems are declining. There's no turning back the clock to an earlier period of robust health. From here on, adaptation will be a primary theme, with lots of work-arounds, prostheses, and propping up to keep things working. We need to give the right kind of care, in the right way. How to proceed? What ethics should guide us? Perhaps the proper model is not far away.

Palliative care and hospice care differ from each other. In the first instance, patients may live a long time, although compromised, and may ultimately die of something unrelated to their debility. In the second, the present illness is deemed terminal, although the time frame for its culmination may be unknown. The differences are important, but in each case the manner of providing care and support is similar. The aim is to alleviate suffering and preserve the highest possible level of awareness, function, and enjoyment. The benefits of such a path accrue to both the patient and the caregivers, as well as to the patient's family and friends.

In the years ahead, the intensifying crises of Planet Earth will elicit a Niagara of crazy behavior. Panaceas will sprout like mushrooms after a rain: "Plant my tree on five million acres and all will be well!" "Build my machine and make carbon pollution a memory!" Myths of denial will morph into myths of escape, and cultish "End Times" fervor will flourish. Some of the hopeful say that, when things get bad enough, the dead-end nonbelievers will finally come to their senses. Let's hope so. Some might shift allegiance and choose to live in a fact-based world. But millions of others will keep walking down Crazy Street because, for them, deviation would cost too much in bewilderment, lost relationships, and eroded identity. People will keep behaving at least as badly, and as well, as people typically behave. Those among us who escape the worst of the calamities will find it challenging to continue necessary work and avoid shutting down.

Which is why the ethics of hospice and palliative care deserve consideration. Prioritize care over cure. Remain unattached to outcomes. Focus on the now. Maintain endurance for the long haul. Such ideas offer an emotionally and spiritually resilient approach to Earthcare. What they imply in terms of projects and policy, however, is difficult to say. A "managed retreat" before the inevitability of sea-level rise, rather than the construction of doomed seawalls, would be consistent with those values, as would material support for those whose homes and cherished places are abandoned in the retreat. Also consistent would be a deep skepticism about "heroic" interventions, such as atmospheric umbrellas and genetically manipulated plants and animals, which may spawn hosts of unintended consequences. Rich societies will have multiple options, poor societies fewer ones. In either case, blanket policies are sure to fray under the wearing realities of cost, societal will, and the peculiarities of individual situations.

My personal, albeit limited, experience in hospice work has taught me that, if you keep your head and are reasonably diligent, surprisingly good things can happen, both for you and for the patient. That word again: "surprise." Everyone has heard the adage, often attributed to Mark Twain, that nine-tenths of good luck is preparation. My guess—call it my hope—is that this principle applies also to Earthcare. Luck favors the prepared planet. It also favors the prepared caregiver.

Along with the toothbrushes, dehydrated eggs, and dried fruit that I lugged to Nepal, there was a bag of several hundred stretchy red rubber bracelets, with "G.R.A.C.E.," in yellow letters, stamped into the band. We distributed them among the members of the expedition and handed them out, along with solar lights and spare clothing, to patients and their loved ones at our clinics.

G.R.A.C.E. is a mnemonic. It stands for steps in a methodology that Roshi Joan, with various colleagues, developed for clinicians. Like most mnemonics, its use of language is a little tortured, but its purpose is pedagogical, not poetic. Its intent is to prompt those who take care of the seriously ill to remember, not so much what to do, but how to be.

G stands for *gathering attention* (the caregiver's, not the patient's).

R for *recalling intentions* (what is the present purpose?).

A for *attuning* (both to oneself and to the patient and others).

C for *considering* (what action will serve the purpose?).

E for *engaging* (doing it) and then *ending* (acknowledging and learning from what has transpired).

It's no accident that four of the five steps precede the actual doing. Preparation. Clarity of mind. The heightened awareness and extra moment it takes to tilt the odds toward benefit. In my first-aid kit, I carry a plastic card that outlines a similar series of steps for assessing emergencies and making first-order diagnoses. When people are in pain, bleeding, or unconscious, your mind speeds up and a hundred thoughts compete for attention. Or maybe you freeze. Unless you are so experienced that reflex alone will produce correct results, a checklist merits the time it takes to consult it. You sort things out; you do things in order; you don't skip steps. G.R.A.C.E. is like that. It helps a caregiver navigate stormy emotional seas. It outlines a process for getting centered and remaining so. The patient receives coherent treatment, and the caregiver benefits too: anxiety and internal conflict might not be banished, but they don't overwhelm.

Some emergencies, like car accidents, happen fast. Others, like climate change and the woes of Petri Earth, begin invisibly and seem to have no end. I keep my G.R.A.C.E. bracelet where I can see it when I work at my desk. Sometimes it helps. It reminds me to approach difficult situations as though they were patients in a clinic: take a deep breath and start at *G*. Get centered. Strong back, soft front. And go from there.

57

Rina

A THIN line, miles away, scars the breast of a mountain, as though long ago a thin blade had sliced the land. Pixels of red, yellow, and black stipple the line, and if you look away and then look back you see that the pixels have moved. They are people in smudged jackets walking a trail. Also yaks bearing grain and lumber. And horses and mules. They are Dolpo afoot, Dolpo in the distance, Dolpo clinging to its vertical world. Above the trail, the top of the mountain is lost in clouds; below, its base lies beyond the reach of the sun. In an hour, maybe two, our train of walkers and mules will reach the mountainside where the pixels inch along. We will follow the scalpel-cut trail to a river foaming down from a pass.

Across the river sprawls a village. The windows of the houses squint darkly at fields, livestock pens, and low rock walls. The roofs, like the slopes behind the village, quiver with flags. Smoke funnels from the rooftops, and the wind snatches it away.

We arrive. A boy with a stick chases a hoop down a lane. Three girls in sooty rags lean together in conference. Suddenly they turn, gap-toothed, eyes bright. "Namaste!" they sing in piccolo voices.

We camp at this village, which is every village: Ringmo, Tokyu, Tinje, Shimen, Koman, Saldang, and more. We are eyed cautiously, then merrily. After a clinic, comes the cultural performance, a thank you for the ibuprofen and amoxicillin, the acupuncture and clean feet, the earnest listening of interpreters and medicos with stethoscopes around their necks. A man comes forward with a dranyan. Six women form a line and chant to its drone, their voices like a chorus of sparrows. They dance in minute, precise steps. They repeat a step three times. Then shuffle. The next step mirrors the first but begins with the other foot. We try to copy the movements. It looks easy. But none of us can do it. A fugitive half beat or quarter step is always lacking. The dance is like culture: you have to learn it from the inside.

The songs wear on; each starts like a dirge but ends at a gallop. We stand in the darkening cold. After two songs we think more about our sleeping bags than we do about the music. After three we say, thank you so much, one more song and we will bid you goodnight. But no, we are told. That would be unwise. To do one more would make the total four, and four of anything is unlucky. Five is no better. There must be six. Well, of course. Thank you. We are tired, but it is all right. We stand in a cold shower of starlight, as the dranyan and the sparrows sing on.

Later, in my tent, I cannot sleep. I switch on a headlamp and rummage in my duffel. I have brought a packet of clippings, items collected in advance of the trip and stored against a night like this. I have read the packet through more than once: the review of a novel by a writer I admire, an essay on the languages of animals, and—what is this? I unfold a wad of newsprint to discover a eulogy for a Pueblo Indian scholar who has died. Her name was Rina Swentzell.

I knew her slightly. She was modest and learned, and she possessed a voice so soft it drew you to the edge of your chair. When I once heard Jane Goodall speak, I thought of Rina. They were cast from the same mold: strong, brilliant women. Rina was Tewa, from Santa Clara Pueblo. In many a presentation, usually before small groups, she would tell her listeners about *p'owaha*, the water-wind-breath that flows through trees, rocks, people, and all things. The Tewa say it is the force that animates the world. In Rina's description, it sounded like the Tao of ancient China, or at least its cousin. She said many of our problems arise when we forget that we share p'owaha with the rest of creation. The invention of anthropomorphic gods flattered us to think we were godlike. Our narcissism misled us into "making the world smaller and smaller until it is nothing but us." I read again the words I underlined in the clipping: "Just human beings out of our natural context. Out of our cosmological context. We have become so small in our view of the world, our world is simply us—human beings."

I put away the clipping and extinguish the headlamp. The silence of the night deepens. Some distance away, a sleeper snores. From farther down the hill comes the cheery singing of a tentful of guides and kitchen staff who seem never to tire and rarely to sleep. Periodically, too, I hear the oohs and ahs of a group of round-eyes who have stayed up to watch for meteors. Drowsy now, I wonder what kind of water-wind-breath reaches that far, into the heavens, and flows in the light of the stars.

58

Phoksundo, Again

WE ARE camped, as before, on the shore of the cerulean lake. The view is oddly sterile. Hardly a breath of wind stirs, and the lake is as flat as glass. No fish, nor swimming or diving bird, nor so much as an errant floating stick animates Phoksundo's jewel-like waters. From across the ravine where the outflow of the lake rushes toward the Suli Gad, stolid Ringmo watches us, its flags queerly limp. We have spread the medical gear on a tarp and are sorting and packing it for the last time, making inventories of the supplies that remain, while noting what we should have brought more of, or less. Not much is left. The "Gyn" bag has been stuffed into the "Gut" bag. And "Wounds" and "Lung" have been interred in "General." The mass of equipment and supplies, once a chest-high mound, would fit in a pair of wheelbarrows.

I have also tried to pack up what I have learned, or think I have learned, but my metaphorical pile is a meager, stingy anthill. I tell myself that, after five weeks and three hundred thousand steps, fatigue has deflated me and things will look better when I am rested, but I don't believe my own reassurance. Deep down, I am pretty sure that my "haul" is as puny as it seems. The sun shines brightly enough. Finches whisper in the juniper scrub. With a clap of wings, a flight of pigeons launches from the roof of the police quarters. But the sunlight has no warmth, and the day feels stale.

This morning I wrote down what I thought were the keys to the kind of medicine we'd been practicing. I had come on the trip thinking they might offer keys to a sane kind of Earthcare, one that preserved its practitioners as well as the systems and creatures they strove to protect. I did not get very far: *Avoid attachment to outcome. Prioritize care over cure. Optimize the present.* It seems a paltry list.

I am wearing the rubber bracelet with "G.R.A.C.E." on it. "Okay," I think, "nice words, good program." But the mnemonic seems personally foreign to me. It isn't my language.

Another addition to my anthill of takeaways is Amchi Lhundup's statement that problems rooted in past lives make a quarter of human illnesses incurable. The notion seems both alien and attractive. Perhaps indeed our species is captive to its accumulation of past lives. Perhaps the long genesis of human nature could have led nowhere else except to the clamoring, possibly incurable, melee of unrestraint now tipping the world into environmental freefall. But where might such an understanding take us, to despair, to forgiveness, to some state where heartbreak and wisdom dwell together? The answer lies beyond my grasp.

Hospice is a compassionate triage. So is Earthcare. You balance a dedication never to quit with the discipline to recognize endings. You pause to honor the passing of whatever is lost: mobility, autonomy, life itself; wildness, species, ecosystems. You remind yourself that, even as Kanjiroba melts, it inspires; that beauty, though diminished, remains. Sometimes you need a lot of reminding.

Packing my gear in Ringmo is hard, not least because of something I cannot unpack: regret. It is the heaviest item in my kit, the leaden mass of all I have not done. In the course of our journey I could have learned, seen, recorded more. I mastered a mere few phrases of Tibetan, never absorbed the stories of so many of the people I met along the way or walked beside, never filled my hours with as much exploration as the mountains and villages invited. I was too often too tired or too lazy, too thirsty for the next cup of tea. The little I can show for the two hundred forty kilometers traversed and the miles of altitude gained and lost is a few notebooks of jottings and a short list of ethics. My anthill. I have collected odds and ends, but I hardly know how to use them. A list is not a practice. I have come a great distance, and somehow hardly moved.

59

Bodhisattva

DOWN FROM the waters of the improbable blue lake, down switchbacks descending the collapsed mountain. Down past the gossamer waterfall, and down to the churning, musical Suli Gad.

We follow the river that is as beautiful as any on Earth, past pines like the pines of home, and past kinnikinnick, wild rose, cinquefoil, juniper, and other plants as familiar as the plants of home.

We walk in seductive forest shade. We hop stones across seeps where pale blue butterflies flurry and sip. The air thickens with the odor of needle duff and forest mold. The trail, suddenly gentle, grows hypnotic. We stand aside when the clip-clopping mule trains overtake us, bells and harness jingling, and when we start again, we go slowly, lulled by the sleepy canyon and the low, rumbling song of the river.

For five weeks we have heard no airplanes. We have inhabited a world powered by muscle and heated by sunlight and dung. These are our last hours to savor it and to crystallize its lessons, yet my lazy thoughts are already drifting homeward. I tell myself to fight them, to remain in the present, but I cannot shake the mental image of settling into a cramped airplane seat for the long flight home, which in this moment seems the epitome of comfort. And that is only one of many hovering delights that now seem so close I can almost smell them: the shower, the sheets, the wine, the massage in the guesthouse basement. Even more, I feel the stirring of a part of me that I have kept in suspension these many days, an almost separate self who now craves messages from home and news of friends and loved ones. I also feel the magnetic pull of the internet and the damnable, addictive news. I want a multicourse meal of world events, with a dessert of sports, trivia, and the doings of the famous. I am a mule ravenous for his nose bag and ration of grain, except that my bag is as fat as Santa's sack and crammed with a thousand shiny things.

I am going down. Down the trail and down the Suli Gad. Down from the heights of Upper Dolpo and down also into old habits and ways, into old mind. I feel the easy allure of passivity, of consumption, of surrender. I try to visualize the gabbling cranes of Namgung and the strong, shining face of the woman whose feet I washed in Tinje, but the images are dim. I try to conjure electric vistas from the crest of Kang La, but this memory, too, seems as flat as a photograph. The immediacy of wild air and the vertiginous closeness of the sky have drained away. A weary apathy heavies my mind and my limbs, and I wonder what, if anything, I can take from the receding past that will retain a breath of life.

Suddenly, noise erupts somewhere ahead. A commotion down the trail. A mule train is coming our way. Here they are: a half-dozen mules bound up-canyon and heavily burdened, one of them carrying an outlandish piece of furniture. The contraption sways wildly past, overmatching its mule and almost slamming the trees. It looks like an immense chair, a Brobdingnagian throne.

Now come the people who belong to those belongings, one of them a vortex of energy and greeting. He stops full in the trail, beaming as though this encounter were the purpose and fulfillment of his day. He is neither short nor tall but large with presence. He wears round, thick-framed eyeglasses, and his face is also round. He has the thinnest of beards, a floppy hat, and forward lean. Name is Phurbu. He laughs hello. Sticks out a hand. Explains that he's a dentist. The contraption that passed on muleback is his patients' chair. It probably won't be last the trip, he says. Ha-ha! Our conversation is brief but leaves an impression. And things won't end there.

Farther up the trail, Phurbu meets Roshi on her horse, with Buddhi leading. She, too, is struck by the dentist's Kris Kringle charm. She peppers him with questions: who is he, where is he going, and why? Turns out he's involved in yet another, albeit smaller, medical expedition to Dolpo. She likes him, and so she recruits him, on the spot, to join future Nomads journeys. She has long wanted a dentist among her throng. Intrigued and amused, Phurbu agrees. Eventually his remarkable biography, teased from many conversations, becomes the talk of the trail and the mess tent.

His story begins with his father's bad leg. Lameness can doom a subsistence farmer. It prevented Phurbu's father from working as productively as other men did, and so his mother did a man's work besides bearing and raising seven children. But even that was not enough. Her double labors provided no more than a stinted existence, and the family remained the poorest of the poor in their Gorkha village. Phurbu was still a child, not yet nine, when his parents sent him and his older brother to Kathmandu to become monks. Now there were two fewer mouths to feed. But as monks the two boys could do little to help their family. Eventually, his brother concluded that for the future well-being of everyone Phurbu must get an education. For a destitute, ethnically Tibetan child like Phurbu, the best opportunity was the Tibetan Children's Villages (TCV) in Dharamsala, India. The Dalai Lama and his sister had founded the school after the Chinese takeover of Tibet. First priority for admission went to actual refugees, second to the children or kinfolk of refugees, and third to the desperately poor. Phurbu qualified only for the third and lowest priority. Nevertheless, an application was made on his behalf. The Village Development Committee back in Gorkha wrote in support. So did a Tibetan refugee committee in Kathmandu. And so did a refugee friend of his brother's. Against steep odds, Phurbu was accepted.

At TCV, almost a decade passed before he saw his parents again, but he says he was never homesick. His classes excited him, and he was ravenous to learn. He soon skipped two or three grades and, even so, ranked first in his class. His brother, meanwhile, also left the insular life of the monastery, cajoling and badgering his way through layers of bureaucracy until he obtained a visa that allowed him to emigrate to New York. He learned to drive a car and got a cabbie's license. He might have gotten married on the money he earned driving taxis, but instead he sent his spare dollars home to sustain the family in Gorkha and to support Phurbu at TCV.

When at last Phurbu graduated, he reunited with his parents in Kathmandu. The sight of his father shocked him. The old man was frighteningly gaunt and his cheeks were sunken. As a child, Phurbu had often seen his father in pain, complaining of toothache. Now he had lost nearly all his teeth. Phurbu said, "You can hardly eat; you need new teeth; you must get dentures." His father said, "Impossible—we can feed the family a long time for the cost of those things." But Phurbu would not relent: his father's health, even his life, depended on his ability to eat. He kept up his campaign of persuasion. Ultimately his father gave in, subject to a condition. He agreed to accept dentures, but only if Phurbu made them.

Phurbu had intended to study medicine. Now he dropped that goal and switched to dentistry. He enrolled in a five-year program in

Bangalore, India. The Nepali government paid one-third of the cost; his long-suffering, cab-driving brother funded the rest. Even before he graduated in 2014, he made dentures for his father.

When we met Phurbu on the trail beside the Suli Gad, he was employed by a clinic in Kathmandu where the senior dentists set rigid prices. They prohibited Phurbu from accepting patients who could not pay for his services. Still, on his own time, Phurbu traveled to villages like the one where he grew up and not only pulled teeth but took molds for dentures that he later fashioned back in Kathmandu. His patients came to collect them in the winter.

Soon that changed. A year or so after his chance meeting with Roshi beside the Suli Gad, Phurbu opened his own clinic in Kathmandu, financed from his meager savings, from loans provided by friends, and from a substantial, no-strings gift of cash from Roshi. At the new clinic, Phurbu calculates fees on a sliding scale, often charging nothing at all. And in the autumn, he joins the Nomads on the trail.

60

On the Trail (2)

PHURBU BLEW by like a fair wind.

Then dull normalcy returned. The hypnotic trail became a trudge. The canyon narrowed, forcing a succession of wearying climbs and longer, jarring descents. I saw myself as a hamster on a slow wheel. I moved my feet robotically, while the numbing hours and the river flowed past. I wish I could tell you that what dawned on me that day came in a blinding, dramatic flash, but it didn't. It seeped into consciousness, like rain into a tent.

I couldn't shake the regret I had felt at Phoksundo, the sense that the meaning of our journey had passed me by. I'd come on the trip thinking that hospice ethics might usefully apply to Earthcare, and nothing I had learned from Roshi, the medicos, or our stoical patients contradicted that. Yet I had not gone further. I had a sense of *so what?* The formulations I had made and the formulations I had borrowed seemed worthwhile, but they were at best only recipes. Where was the real grub?

I dwelled in such ruminations, plodding along, when I became aware of a kind of counterpoint welling up in the background of my thoughts. No one can explain how words and ideas come to mind. The manner of their summoning eludes us. Yet the mystery of unframed thoughts becoming manifest attends us daily, and every walker knows that walking seems to foster it. It is as though the physical shock of our footfalls shakes loose notions we never suspected our minds contained and frees them to float into consciousness.

At first, I couldn't grasp the idea hovering in the back of my mind, couldn't cast it as a thought. It was a shadow, no more. Then, strange to say, a light fell on it and it took form. It presented itself as a sentence: "The right way to carry the grief is the right way of walking." I heard it in my mind's ear, as though it were made of sound. I replayed it, hearing it again. Once. Twice. Several times. The words were clear but the sentence seemed a riddle.

I was walking. The sentence had arrived like a telegram, sender unknown. It had a rhythm that fit my steps. I marched to it, silently chanting. I didn't understand what the words were trying to tell me, but I liked their sound and I kept their cadence. I walked a hundred paces, repeating the sentence like a mantra. Then I reversed the phrases and walked another hundred: "The right way of walking is the right way to carry the grief."

When I thought about the grief, it was not hard to identify. It was the stew of sorrows brewed from climate change and the woes of Petri Earth. It was the distress aroused by our planetary dilemma and by our failure as a society, even as a species, to respond in full, to assent to what we know.

Grief was not the riddle. Walking was. What was the right way? Our expedition had walked for five weeks. Life had contracted to the essentials of food and shelter, movement and rest, work and play. Every hour's effort had been embedded in a stern geography. The requirements of place and mission asserted demands that none of us could meet alone. We had to cooperate, act together, and so we formed a community. We served. We learned. We walked. We found a settled pace, literally and figuratively. For a time, we were nomads. The trail became home.

Tenzin had said, "Every day a yatra"—a pilgrimage. And so it felt: each day with a new destination or new people to treat; each day with an obligation to carry ourselves with composure and care; each day requiring response to what it offered: foul weather, obstacles, accident—any situation that might arise. I thought back to our encounter with the mule with the broken leg. I remembered—and lamented—the paralysis I felt at the sight of it. And then recalled how the two women, the doctor and the nurse, responded. It was plain to all that the mule must die, but what they knew and I had failed to grasp was that the mule did not have to die in misery. For them the creature was no different from a patient in a clinic. They gave it comfort; they gave it treatment, the best they could. And more comfort as the treatment took hold. Their backs were strong, their fronts soft. They walked the right way.

In one day more, the trail beside the Suli Gad would end. Each day beyond that would be another trail, another yatra, in cities and towns, deserts and mountains. The lesson had been obvious all along: there is no place that is not the trail, no journey that is not—or should not be—a yatra. When you are making a yatra, you cannot be invisible, cannot fail to show up. Futility and resignation have no place. On a yatra every step and action matter, and an act well done leaves something under the skin, a higher level of animation, a firmer kind of belief. Perhaps the most important thing it imparts is an appetite for doing more. The voyage of the *Beagle* was Darwin's pilgrimage, and he was still on the trail, sick but persevering, at his writing desk at Down. Wegener, convalescing from war wounds or on the ice in Greenland, was no less a pilgrim, ever striving toward a fuller vision of the genesis of the world, and so he kept revising his cherished book. The same steadfastness calls to you and me, to all of us: we are voyagers through time, pilgrims in motion. Or rather, we can be. Perhaps we must be if we are to espouse the species of hope that maintains its readiness for surprise.

How to walk? The answer abides in the trail, in the doing. The trail had taught me that I still had far to go in learning to walk. But it showed the way. The trail was not separate from the expedition and its values, from the daily connectedness, the camaraderie of shared tasks, the pacing, the self-care, the meditative rhythms, the silences. And it was not separate from the work we did or from the meaning that derived from the work. The trail led us to the villagers who were our hosts and our resilient patients, who guided us by example. In our time with them, we may have saved some lives, we may have saved our own—perhaps one saving includes the other. And always, all around us, the land presided. It contained our traveling and our living. It immersed us in an immense, austere beauty that was at once impermanent and eternal, thrilling and stern. Had some of its spaciousness seeped into me? Did I now carry traces of its spirit in my bones and blood? I prayed so. Muscle memory survives when details of sight and sound begin to fade. The deep memory of our bodies gives us balance, like the balance that allows us to ride a bicycle, a balance we never entirely forget.

"All we have been through," Tenzin had said, "you will feel lighter." And I did. When I stopped thinking about the distance we had come and the distance yet to go, I felt the simple joy of motion. Motion toward the next camp, the next situation, the next task, the next ark to be built. No matter if I fall off my yatra bike, which I surely will, I will get back on.

61

Kora

AT BOUDHA Stupa in Kathmandu the kora never ends. A river of people circles the great round temple, day and night, day after day. The dome of the stupa, topped by a chorten-like tower bearing the inevitable watchful eyes, swells above the skyline of the city, and beneath its gaze humanity swirls. Monks, pilgrims, and the merely curious light butter lamps, make offerings, beg, pray, and chant. They walk cumulative millions of laps around the giant structure, hundreds of walkers in any daylight hour, so that the surrounding plaza seems to spin with people, and the motion, playing its trick on the eye, appears to cause the massive shrine to turn.

In generations past, the stupa stood alone in a bucolic valley outside the city. But now Kathmandu, ten times its size of sixty years ago, has swallowed it. Monasteries, teashops, and stalls crammed with religious knickknacks press close upon the temple, the million-person metropolis leaning in from every side.

Monks chant from doorways and benches. They squat on the brick pavement, bobbing their heads in constant repetition, holding out their begging bowls to the circular current of walkers. Many ordinary people log three quick circuits of the stupa en route to work and then do the same at the end of the day, earning merit for the next life. They join a stream of humanity thick with bead counters and liturgy chanters, who mumble with breathless incessancy. The throng swerves like water around penitents with wooden paddles on their hands who clap them down and scrape forward to lie prostrate on the filthy ground, then mumble a mantra, push up, take three or six steps, and clap down again. The one-legged and the one-armed are also here, their bandages sometimes leaking from new stumps. A doe-eyed child with miniature, useless legs pushes along on a skateboard, pleading for money. He skims past a man who stares into infinity from empty eye sockets, a bewildered half smile on his scarred face, as he, too, waits for alms.

Sleeping dogs, like boulders, block the human stream, the crowd parting around them. Suddenly two curs come to life and lunge snarling at each other, their anger brief and inconclusive. From an alley a line of shaven-headed monks in russet robes shuffles forth, joining the current like a file of bald ducklings.

Walkers closest to the shrine spin the prayer wheels housed in niches in the stupa walls. The wheels clatter and clack. The padding and scuffing of feet raise a murmur from the bricks, like a fall of rain. The paddles of the inch-worming penitents clap and scrape. From nearby gompas come sudden crescendos of cymbals and flatulent horns. The human carousel continues, past ranks of flickering butter lamps, through clouds of sickly incense.

In the widest part of the plaza, old women preside over plastic bowls brimming with corn. They sell the grain to those who would earn merit by feeding the pigeons that graze upon the moss or algae growing on the dome of the stupa, which is said to weaken the plaster surface. The pigeons and those who feed them thus come to the aid of the holy structure, except that the corn provided is more than any army of welfare pigeons can eat, and so the dome goes unpecked.

Suddenly a helicopter grates the sky, lifting the pigeons in a flapping cloud. Below them, the sick, the wounded, the daily commuters, the weak and the well, the wheel spinners and mantra mumblers march on. The dogs squall, the monks beg, and the tourists sip their coffees looking on. The kora revolves, old and young, rich and poor, many with masks against sickness and pollution, the entire clockwise rotation of humanity's parade turning and turning, like a child's top, like a prayer wheel, like a spinning planet.

Every day a yatra
Every situation a clinic
Absorb the beauty
Build an ark
Be alive

A Note on Place-Names and Language

THIS BOOK is set in Dolpo, but many published maps, especially those originating from Kathmandu, call the district Dolpa. The former term is Tibetan, the latter Nepali. Other variations are not so easy to explain. Phoksumdo, with an *m*, may lie closer to the name's Tibetan roots, but Phoksundo, with an *n*, prevails in most contexts. The same is true of Kanjiroba, which appears interchangeable with Kanjirowa, and some maps show multiple mountains bearing the same name. Still more problematic is the rendering of yarza gunbu, the caterpillar-borne fungus and health cure so eagerly sought from Dolpo's high-altitude grasslands. Such is the difficulty of transposing regional dialects of Tibetan (for the substance is known from multiple language areas) that variations of its spelling in written English seem almost limitless.

A SHORT GLOSSARY OF TERMS USED IN THE TEXT

Bharal, also **bhara**, blue sheep, *Pseudois nayaur*

Bodhisattva, a person on the path toward enlightenment

Bön, also **B'on** or **Bon**, a Tibetan religion distinct from Tibetan Buddhism but espousing similar values and practices

Bön-pos, followers of Bön

Boudha Stupa, also **Boudhanath**, one of the world's largest stupas, eleven kilometers from the center of Kathmandu, a UNESCO World Heritage site

Chorten, a monument or shrine containing religious relics, possibly including the ashes of a noted monk or religious teacher; the Tibetan analog of the Sanskrit term "stupa"

Dolpo-pa, the people of Dolpo

Dranyan, a stringed musical instrument, "the Himalayan lute"

Gompa, a place of refuge, a monastery, a convent, or the temple associated with such places

Himalaya, the vast assemblage of mountain ranges known as the "abode of snow"; its devotees avoid referring to it in the plural as "the Himalayas"

Kailash, also **Kailas**, a 6,638-meter peak in southwestern Tibet where four of Asia's greatest rivers—the Indus, Brahmaputra, Sutlej, and Karnali (tributary to the Ganges)—have their source. The mountain is regarded as sacred in Tibetan Buddhism, Bön, Jainism, and Hinduism

Kora, a circular pilgrimage

Lama, a monk who has completed a three-year retreat emphasizing silence and meditation

Netsang, a system of trade kinship in which families unrelated by blood come to regard each other as kin based on a sustained trading partnership generations deep

Pecha, a sacred book or document

Puja, a religious ceremony

Sherpa, an ethnic group in northeastern Nepal, where Mount Everest is located. Because people from that group have long served as highly skilled guides on mountaineering expeditions to Everest and other major summits, the term "Sherpa" has been applied to guides from other ethnic groups who perform a similar function

Thobo, a pile of stones, regarded as a shrine, marking the top of a mountain pass or some other noteworthy point in the landscape, to which a pilgrim may add a stone out of respect and gratitude for arriving there. *Thobo* is commonly used in Dolpo and perhaps surrounding areas. It is synonymous with *doubeng*, which is used elsewhere in Nepal and is more familiar to most visitors to the region

Yarza gunbu, also **yartsa gumba** and many other spellings, which translates as "summer grass, winter worm," a fungus, *Ophiocordyceps sinensis*, that parasitizes the caterpillar larvae of ghost moths (*Hepialus humuli*) at high altitudes in the Himalaya and the Tibetan Plateau; its consumption is considered to be salutary for health in general and for male virility in particular

Yatra, a journey with religious or spiritual intent, a pilgrimage

Yeti, the legendary wild hominid of the Himalaya, the "Abominable Snowman"

Sources and Suggested Readings

Introduction

David Quammen, *Spillover: Animal Infections and the Next Human Pandemic* (New York: Norton, 2012), 21, 23.

1 ✦ First Things

Yuval Noah Harari, *Sapiens: A Brief History of Humankind* (New York: HarperCollins, 2015).

Robert M. Hazen, *The Story of Earth: The First 4.5 Billion Years, from Stardust to Living Planet* (New York: Penguin, 2012, 2013).

Elizabeth Kolbert, *The Sixth Extinction: An Unnatural History* (New York: Henry Holt, 2014).

3 ✦ Levels of Amazement

Footprintnetwork.org.

Vaclav Smil, "Harvesting the Biosphere: The Human Impact," *Population and Development Review* 37, no. 4 (December 2011): 613–36.

4 ✦ The Suli Gad

Peter Matthiessen, *The Snow Leopard* (New York: Penguin, 1978, 2008), 127, 131.

George B. Schaller, *Stones of Silence: Journeys in the Himalaya* (Chicago: University of Chicago Press, 1980).

5 ✦ A Microdot of the Great Rent

Henry Frankel, "Continental Drift and Plate Tectonics," in *Sciences of the Earth: An Encyclopedia of Events, People, and Phenomena*, ed. Gregory A. Good (New York: Garland Publishing, 1998).

David M. Richardson, ed., *Ecology and Biogeography of Pinus* (Cambridge, UK: Cambridge University Press, 2000).

6 ✦ Circle

Concerning Mount Kailash: see Maharaj K. Pandit, *Life in the Himalaya: An Ecosystem at Risk* (Cambridge, MA: Harvard University Press, 2017), 45.

7 ✦ There Is No Problem Here

David A. Brew, "Preliminary Report on Geologic Features of Shey-Phoksundo National Park, Dolpa, Nepal," US Geological Survey Open File Report 91–117, prepared in cooperation with the Kingdom of Nepal, Department of National Parks and Wildlife Conservation, 1991.

Phillip Sturgeon, "Reflections on Lake Phoksumdo," *The Himalayan Journal* 55, 1999. https://www.himalayanclub.org/hj/55/5/ reflections-on-lake-phoksumdo/.

8 ✦ Dramatis Personae

Information on Action Dolpo, founded by Marie-Claire Gentric, may be found at actiondolpo.com.

Abraham Verghese, *Cutting for Stone* (New York: Vintage Books, 2009), 7.

9 ✦ Uplift

Charles Darwin, *The Origin of Species and The Voyage of the Beagle* (New York: Alfred A. Knopf, 2003), 313, 314, 321, and chapter 15, *passim*.

Adrian Desmond and James Moore, *Darwin: The Life of a Tormented Evolutionist* (New York: Norton, 1991), 108 (Henslow quoted, "on no account").

George Eliot, *Middlemarch* (New York: Random House, 1984), 591.

Frankel, "Continental Drift and Plate Tectonics."

Richard Nelsson, "Mount Everest: Named After the First Surveyor General of India," *The Guardian*, July 21, 2011, https://www. theguardian.com/theguardian/from-the-archive-blog/2011/jul/21/ mount-everest-name-1856/.

Stewart Green, "The Geology of Mount Everest," LiveAbout, May 18, 2018. https://www.liveabout.com/geology-of-mount-everest-755308/.

Binaj Gurubacharya, "China, Nepal Say Everest a Bit Higher than Past Measurements," *Washington Post*, December 8, 2020, https://www. washingtonpost.com/world/asia_pacific/china-nepal-say-everest-a-bit-higher-than-past-measurements/2020/12/08/3a63b738-3933-11eb-aad9-8959227280c4_story.html.

10 ✦ Drift

H. E. Le Grand, *Drifting Continents and Shifting Theories* (Cambridge, UK: Cambridge University Press, 1988).

Naomi Oreskes, *The Rejection of Continental Drift: Theory and Method in American Earth Science* (New York: Oxford University Press, 1999).

Alfred Wegener, *The Origin of Continents and Oceans*, trans. John Biram (New York: Dover: 1929, 1966).

11 ✦ Bogu La, 16,959 feet

Gerald Wiener, Han Jianlin, and Long Ruijun, *The Yak*, 2nd ed., RAP publication 2003/06 (Bangkok: Regional Office for Asia and the Pacific, Food and Agriculture Organization of the United Nations, 2003).

Mountain Research Initiative EDW Working Group, "Elevation-Dependent Warming in Mountain Regions of the World," *Nature Climate Change* 5, (2015): 424–30, https://doi.org/10.1038/nclimate2563/.

14 ✦ Roshi

Atul Gawande, *Being Mortal* (New York: Henry Holt, 2014).

Joan Halifax, *The Fruitful Darkness* (HarperSanFrancisco, 1993).

The Darwin quotation is from the final sentence of *Origin of Species*, which is discussed in detail in chapter 28, "The Grandeur Sentence."

15 ✦ Hospice for a Mouse

Daniel Southerl, "China Said to Suppress Tibet's Buddhism," *Washington Post*, September 21, 1990.

The International Campaign for Tibet, "When the Sky Fell to Earth: The New Crackdown on Buddhism in Tibet," Washington, DC, 2004.

16 ✦ Lumber Yaks

Kenneth M. Bauer, *High Frontiers: Dolpo and the Changing World of Himalayan Pastoralists* (New York: Columbia University Press, 2004).

17 ✦ Aunts and Uncles

Walt Whitman, "Song of the Open Road" and "Crossing Brooklyn Ferry," *Leaves of Grass and Selected Prose* (New York: Modern Library, 1921,1950), 118, 128. Originally published 1856.

18 ✦ Hidden Figure

Erin Blakemore, "Seeing Is Believing: How Marie Tharp Changed Geology Forever," *Smithsonian Magazine*, August 30, 2016, https://www.smithsonianmag.com/history/seeing-believing-how-marie-tharp-changed-geology-forever-180960192/.

Hali Felt, *Soundings: The Story of the Remarkable Woman Who Mapped the Ocean Floor* (New York: Henry Holt, 2014).

19 ✦ Geopoetry

William Glen, *The Road to Jaramillo: Critical Years of the Revolution in Earth Science* (Palo Alto: Stanford University Press, 1982).

Harold L. James, "Harry Hammond Hess, 1906–1969," in *Biographical Memoirs*, National Academy of Sciences, 1973, http://www.nasonline.org/publications/biographical-memoirs/memoir-pdfs/hess-harry.pdf.

H. E. Le Grand, *Drifting Continents and Shifting Theories* (Cambridge, UK: Cambridge University Press, 1988).

John Tuzo Wilson, "A Possible Origin of the Hawaiian Islands," *Canadian Journal of Physics* 41, no. 6 (June 1963): 863–70, https://doi.org/10.1139/p63-94.

21 ✦ Feet

Joan Halifax, *Standing at the Edge* (New York: Flatiron Books, 2018), 138–39.

22 ✦ Tinje ЯUs

Concerning Olafur Eliasson and Little Sun solar lights: https://littlesun.com.

Bauer, *High Frontiers*, 104, concerning Tinje airstrip.

Rebecca Solnit, "Medical Mountaineers," *New Yorker*, December 21 & 28, 2015, 78–87.

Sam Cowan, *Essays on Nepal, Past and Present* (Kathmandu: Himal Books, 2018).

Kanta Kumari Rigaud, Alex de Sherbinin, Bryan Jones, Jonas Bergmann, Viviane Clement, Kayly Ober, Jacob Schewe, Susana Adamo, Brent McCusker, Silke Heuser, and Amelia Midgley, "Groundswell: Preparing for Internal Climate Migration," The World Bank (2018), https://open-knowledge.worldbank.org/handle/10986/29461/.

Mark Lynas, *Our Final Warning: Six Degrees of Climate Emergency* (London: 4th Estate, 2020).

Bill McKibben, "130 Degrees," *New York Review of Books* 57, no. 13 (August 20, 2020): 8–10.

John Podesta, "The Climate Crisis, Migration, and Refugees," Brookings Institution Policy Brief (July 25, 2019), https://www.brookings.edu/research/the-climate-crisis-migration-and-refugees/.

Alex Randall, "Climate Refugees: How Many Are There? How Many Will There Be?," Climate and Migration Coalition, http://climatemigration.org.uk/climate-refugees-how-many/.

24 ✦ Lone Wolf

George B. Schaller, "Comments on the Management of Shey-Phoksundo National Park" (October 24, 2016), MS in author's possession.

25 ✦ An Entire Heaven and an Entire Earth

Gerardo Ceballos, Paul R. Ehrlich, and Rodolfo Dirzo, "Biological Annihilation via the Ongoing Sixth Mass Extinction Signaled by Vertebrate Population Losses and Declines," *Proceedings of the National Academy of Sciences* (July 2017): 1704949114v1-201704949.

Henry David Thoreau, *The Writings of Henry D. Thoreau*, Online Journal Transcripts, http://thoreau.library.ucsb.edu/writings_journals_pdfs/J10f3-f4.pdf. Thanks to Dan Flores for pointing out Thoreau's entry for March 23, 1856, which he quotes in *American Serengeti* (Lawrence: University of Kansas Press, 2016), 9.

Curt Meine, *Aldo Leopold: His Life and Work* (Madison: University of Wisconsin, 1988).

Aldo Leopold, "The Green Lagoons," in *A Sand County Almanac with Essays on Conservation from Round River* (New York: Oxford University Press, 1949, 1968), 148–49.

Denise Lu, "We would need 1.7 Earths to make our consumption sustainable," *Washington Post*, May 4, 2017, https://www.washingtonpost.com/graphics/world/ecological-footprint/?utm_term=.c0495c4b043d.

Aaron Retica, "Kurosawa on Earthquakes," *New York Times*, March 14, 2011, https://6thfloor.blogs.nytimes.com/2011/03/14/kurosawa-on-earthquakes/?_r=0.

Bert Cardullo, ed., *Akira Kurosawa: Interviews (Conversations with Filmmakers)* (Jackson, MS: University of Mississippi Press, 2007).

28 ✦ The Grandeur Sentence

At darwin-online.org.uk, one finds searchable digitized versions of all editions of *On the Origin of Species*.

Charles Darwin, *The Foundations of the Origin of Species, a Sketch Written in 1842*, ed. Francis Darwin (Cambridge: Cambridge University Press, 1909). Francis Darwin, son of Charles, cites a still earlier antecedent for the grandeur sentence in his father's "Note Book of 1837" (xxi–xxii).

Desmond and Moore, *Darwin*. Jane Gray's quote is from p. 562.

Peter Salwen, "Chagas' Disease Claimed an Eminent Victim," *New York Times*, National Edition, Thursday, June 15, 1989, A30.

29 ✦ The Grandeur of Saldang

Darwin, *Voyage of the Beagle*, esp. chapter 10, 216–42.

Charles Darwin, *The Descent of Man, and Selection in Relation to Sex*, (New York and London: Penguin Books, 1879, 2004). (Concerning Tierra del Fuego, see xix–xx.)

30 ✦ Galapagos

Darwin, *The Foundations of the Origin of Species*. See especially the introduction in which Francis Darwin discusses conflicting views on when his father conceived the theory of natural selection.

Desmond and Moore, *Darwin*, 185–86, 214, 328.

Charles Darwin, *The Autobiography of Charles Darwin, 1809–1882*, ed. Nora Barlow (New York: Norton, 1958, 1993), 118–19.

Darwin, *Voyage of the Beagle*, 390, 405–6, 409, 413.

31 ✦ Invisible

Eliot, *Middlemarch*, 189.

32 ✦ Mechanism

Oreskes, *Rejection of Continental Drift*.

Thomas Robert Malthus, *An Essay on the Principle of Population and Other Writings*, ed. Robert Mayhew (London: Penguin Random House, 2015).

33 ✦ Saldang Clinic

Matthiessen, *Snow Leopard*, 268.

34 ✦ Doing the Math

Herman Melville, *Moby-Dick*, eds. Harrison Hayford and Hershel Parker (New York: W.W. Norton, 1967), chapter 96, 354–55.

Darwin, *Voyage of the Beagle*, 410.

William deBuys, *The Last Unicorn: A Search for One of Earth's Rarest Creatures* (New York: Little, Brown and Company, 2015).

35 ✦ Survival of the Sexiest

Darwin, *Descent of Man.*

David Rothenberg, *Survival of the Beautiful: Art, Science, and Evolution* (New York: Bloomsbury, 2011).

James Watson, *The Double Helix: A Personal Account of the Discovery of the Structure of DNA* (New York: Touchstone, 1998), 210. *The Double Helix* was first published in the US by Atheneum in 1968.

36 ✦ The Cranes of Namgung

Project Noah at https://www.projectnoah.org/spottings/8840974.

Guo Yumin and He Fenqi, "Preliminary Results of Satellite Tracking on Ordos Demoiselle Cranes," *Chinese Journal of Wildlife* 38, no. 1 (January 2017): 141–43.

37 ✦ Flowing Mountains

Zen Master Dogen, "Mountains and Waters Discourse (1) (Sansui Kyo)," trans. Kazuaki Tanahashi, Upaya Zen Center, https://www.upaya.org/uploads/pdfs/MountainsRiversSutra.pdf.

George B. Schaller, *Tibet Wild: A Naturalist's Journeys on the Roof of the World*, (Washington, DC: Island Press, 2012).

38 ✦ Flowing Seafloor

Glen, *Road to Jaramillo*, chapter 7.

Frederick J. Vine and Drummond H. Matthews, "Magnetic Anomalies over Ocean Ridges," *Nature* 199 (September 7, 1963): 947–49.

Lawrence W. Morley, "Early Work Leading to the Explanation of the Banded Geomagnetic Imprinting of the Ocean Floor," *EOS* 67, no. 36 (September 9, 1986): 665–66; doi.org/10.1029/EO067i036p00665.

39 ✦ Tuzo Wilson

John Tuzo Wilson, "Continental Drift," *Scientific American* 208, no. 4 (April 1963): 86–103, http://www.jstor.org/stable/24936535.

John Tuzo Wilson, "A New Class of Faults and Their Bearing on Continental Drift," *Nature* 207 (July 24, 1965): 343–47, https://www.nature.com/articles/207343a0.

40 ✦ The Jaramillo Event

Glen, *Road to Jaramillo*, 347.

Fraser Goff, *Valles Caldera: A Geologic History* (Albuquerque: University of New Mexico Press, 2009).

G. Brent Dalrymple, "Richard R. Doell, 1923–2008," in *Biographical Memoirs*, National Academy of Sciences, http://www.nasonline.org/publications/biographical-memoirs/memoir-pdfs/doell-richard.pdf.

Richard R. Doell, G. Brent Dalrymple, Robert L. Smith, and Roy A. Bailey, "Paleomagnetism. Potassium-Argon Ages, and Geology of Rhyolites and Associated Rocks of the Valles Caldera, New Mexico," The Geological Society of America, Memoir 116 (1968).

Ana Steffen, Richard Hughes, Tim Aydelott, and Ylonda Viola, Interview with Robert L. Smith, Sacramento, Calif., November 2011. Courtesy of Ana Steffen, Valles Caldera National Preserve, Jemez Springs, New Mexico.

42 ✦ Prayer Mills
Matthiessen, *Snow Leopard*, 193.
Sean Carroll, *The Big Picture: On the Origins of Life, Meaning, and the Universe Itself* (New York: Dutton, 2016).
Erik Vance, "Mind Over Matter," *National Geographic* 230, no. 6 (December 2016): 30–55.
Erik Vance, *Suggestible You: The Curious Science of Your Brain's Ability to Deceive, Transform, and Heal* (Washington, DC: National Geographic Books, 2016).

45 ✦ Hope
Václav Havel, *Disturbing the Peace: A Conversation with Karel Huizdala* (New York: Vintage, 1991).
Stephen L. Fisher and Barbara Kingsolver, a conversation at Emory & Henry College Literary Festival, September 30, 2011, *Iron Mountain Review* 28 (2012).
Joan Halifax, "Wise Hope in Social Engagement," Upaya Zen Center, December 31, 2018, https://www.upaya.org/2018/12/wise-hope-in-social-engagement-by-roshi-joan-halifax-part-1/.

46 ✦ Tsakang
Matthiessen, *Snow Leopard*, esp. 200–1, 242.

47 ✦ Sacred Rage
Donald S. Lopez Jr., *The Story of Buddhism: A Concise Guide to Its History and Teachings* (New York: HarperCollins, 2001).
Douglas Preston, "I Took the Dalai Lama to a Ski Resort and He Told Me the Meaning of Life," *Slate*, November 16, 2014.
Terry Tempest Williams, *The Hour of Land: A Personal Topography of America's National Parks* (New York: Sarah Crichton Books, 2016).

48 ✦ Kang La
Matthiessen, *Snow Leopard*, 175, 182.

49 ✦ Birches
Pandit, *Life in the Himalaya*.
K. S. Valdiya, *Dynamic Himalaya* (Hyderabad: Universities Press, 1998), esp. 93–95, 121 ff.

50 ✦ Snow Leopard
A very remote possibility was that the tracks were laid down by the smaller Eurasian lynx, which has been reported considerably to the east, near Dhaulagiri, and is otherwise not known from Dolpo.

Kristin Nowell, Juan Li, Mikhail Paltsyn, and Rishi Kumar Sharma, "An Ounce of Prevention: Snow Leopard Crime Revisited," TRAFFIC, Cambridge, UK, October 2016; https://www.traffic.org/site/assets/files/2358/ounce-of-prevention.pdf.

Sandra Diaz, Josef Settele, Eduardo Brondizio co-chairs, and twenty-six others, "Summary for Policymakers of the Global Assessment Report on Biodiversity and Ecosystem Services—Advance Unedited Version," Intergovernmental Science-Policy Platform on Biodiversity and Ecosystem Services, May 6, 2019, https://www.ipbes.net/sites/default/files/downloads/spm_unedited_advance_for_posting_htn.pdf.

Pandit, *Life in the Himalaya*.

Schaller, "Comments on the Management of Shey-Phoksundo National Park."

51 ✦ Amchi

Sienna R. Craig, *Healing Elements: Efficacy and the Social Ecologies of Tibetan Medicine* (Berkeley: University of California Press, 2012).

52 ✦ Yeti

Matthiessen, *Snow Leopard*, 118–20, 133.

53 ✦ Clinic Notes

Matthiessen, *Snow Leopard*, 19.

Halifax, *Standing at the Edge*.

Cynda Rushton, Alfred Kaszniak, and Joan Halifax, "A Framework for Understanding Moral Distress among Palliative Care Clinicians," *Journal of Palliative Medicine* 16, no. 9 (September 16, 2013), https://doi.org/10.1089/jpm.2012.0490.

54 ✦ Kanjiroba

Detailed maps may show peaks confusingly labeled Kanjirowa North and Kanjorowa South sixteen miles to the northwest of the Kanjiroba in the text, but neither of these is the peak that we have been seeing. For maps scaled at 1:50,000 with 40-meter contours, see http://pahar.in/nepal-topo-maps/. Map 2982-16, "Phoksundo Tal," is a good place to start.

Pandit, *Life in the Himalaya*, 221, 251.

Joshua M. Maurer, Summer B. Rupper, and Joerg M. Schaefer, "Quantifying Ice Loss in the Eastern Himalayas since 1974 Using Declassified Spy Satellite Imagery," *The Cryosphere* 10 (2016), 2203–15, https://doi:10.5194/tc-10-2203-2016.

Mountain Research Initiative EDW Working Group, "Elevation-Dependent Warming in Mountain Regions of the World."

Keegan McChesney, "Existential Avalanche: The Lived Experience of Climate Change in Dolpo and Mustang, Nepal," Independent Study Project Collection, Paper 2091 (2015), http://digitalcollections.sit.edu/isp_collection/2091.

F. Scott Fitzgerald, "The Crack-Up," in *Esquire's Big Book of Great Writing*, ed. Adrienne Miller (New York: Hearst Books, 2003). (Originally published in *Esquire Magazine*, February, March, and April 1936.)

55 ✦ Cutting Deals

Peter Ward and Joe Kirschvink, *A New History of Life: The Radical New Discoveries about the Origins and Evolution of Life on Earth* (New York: Bloomsbury, 2015).

Darwin, *Descent of Man*, chapters 3–4, quotes from pp. 147, 151. The story of the monkey defending its keeper is at p. 126.

Richard Powers, *The Overstory* (New York: Norton, 2018).

Peter Wohlleben, *The Hidden Life of Trees: What They Feel, How They Communicate—Discoveries from a Secret World* (Vancouver: Greystone Books, 2016).

56 ✦ Hospice

Chris D. Thomas, *Inheritors of the Earth: How Nature Is Thriving in an Age of Extinction* (New York: PublicAffairs, 2017).

Bill McKibben, *Eaarth: Making a Life on a Tough New Planet* (New York: St. Martin's Griffin, 2010, 2011).

Warren Cornwall, "Heatproofing Coral: Fast-Forwarding Evolution to Save the Reefs," *Science* 363, no. 6433 (March 22, 2019): 1264–69.

Joan Halifax, "G.R.A.C.E. for Nurses: Cultivating Compassion in Nurse/Patient Interactions," *Journal of Nursing Education and Practice* 4, no. 1 (2014): 121–28, http://dx.doi.org/10.5430/jnep.v4n1p121.

57 ✦ Rina

Jack Loeffler, "Honoring Rina," *Green Fire Times*, December 2015, 15–16, 22, 28.

Jack Loeffler, *Survival along the Continental Divide: An Anthology of Interviews* (Albuquerque: University of New Mexico Press, 2008), 85–102 (interview with Rina Swentzell)

Acknowledgments

FIRST THANKS go to Roshi Joan Halifax, who supported this project, start to finish. Her invitation to join Nomads began the journey, and her comments on key portions of the text helped to bring it home. A bow of deepest gratitude.

My thanks also to all the Nomads, both Western and Nepali, comrades on the trail in 2016 and '18 and friends ever after. Only a few are named in the text, but my sincere appreciation and affection extends to each of them. Namaste.

Also essential was Don Lamm, my unofficial agent on this project (and official on several past). Once again, his savvy, generosity, and confidence overcame every impasse.

Additional thanks to Catherine Baca for transcribing—yet again—the scrawl of my field notes, to Beth Hadas and David Gaines for critiques of the manuscript, to Bill Cronon, Sally Butler, and Trevor Smith for ideas they may not know they contributed, to Terry Tempest Williams, Dan Flores, and Doug Preston for ideas they know they did, to Becky Gaal for the wonderful map and drawings, to Ana Steffen for fleshing out the Valles Caldera story, and to Christa Sadler, Marty Peale, and Don Usner for helpful conversations and forays of various kinds. And speaking of forays, immense thanks to Ted and Betsy Rogers and the late and much lamented Phil Smith for the unforgettable experience of the Galapagos.

At Seven Stories Press, Dan Simon, Lauren Hooker, Stewart Cauley, Conor O'Brien, and others helped bring the book into being. My thanks for their efforts, and especially to Bill Rusin, who introduced the book to them and generously shepherded it to final publication.

This book covers a lot of ground in geography, culture, and science, and, although I have done my best to eliminate errors of fact and interpretation, some no doubt remain. They are mine alone.

It may sound precious for me to offer thanks to the people of Dolpo, but I wish to acknowledge that I learned much from them and carry their memory with affection and earnest hope for their future.

Finally, to Joanna, as ever, for invaluable support, thank you and love.

Index

About the Author

WILLIAM deBUYS is the author of ten books, including *The Last Unicorn*, one of *Christian Science Monitor*'s 10 Best Nonfiction Books of 2015; *River of Traps*, a *New York Times* Notable Book of the Year and a Pulitzer Prize nonfiction finalist; *The Walk* (an excerpt of which won a Pushcart Prize in 2008); and *A Great Aridness*. In 2008–2009 he was a Guggenheim Fellow. He lives in New Mexico.